EXECUTE
YOUR
TECH IDEA

A Step-by-Step Guide for Non-Techies,
Professionals, Managers, and Startups

Andrew Ward

Triple Twist Publishing™

Triple Twist Publishing Ltd
13 Portland Rd
Birmingham
B16 9HN

Twitter: @executeyourtech
Facebook: @Executeyourtechidea
LinkedIn: @executeyourtechidea
Instagram: @executeyourtechidea

ExecuteYourTechIdea.com

CONTENTS

EXECUTE YOUR TECH IDEA

Advanced reader testimonials for Execute Your Tech Idea:

"Each chapter builds on the previous one and gradually increases the "techiness" but in a way that allows the reader to grow in confidence. Andrew's gentle and gradual manner of introducing each step and logically moving through the entire process step by step with this very solid structure made the experience enjoyable and very enlightening."

- USA *Today Bestselling Author, Astrid V.J*

"It was very clear to me while reading that Andrew has a deep and extensive knowledge of the technology space, and did very well in translating that knowledge in a digestible and approachable way for readers."

- *Ryan Wood, Executive at Celtx*

"If you are approaching a tech project without any previous knowledge of tech or best business practices regarding a tech project, this is the must-read book. Andrew has masterfully written a book that is easy to read, understand and will give you the full picture of how to execute a project. Do not waste your time consulting with developers until you have read this book and gained a clear understanding and vision!"

- *Marcia Leonard, Entrepreneur*

INTRODUCTION

"You miss 100 percent of the shots you never take."
—Wayne Gretzky, Leading goal scorer in NHL

There was a time when only the most cutting-edge software businesses built tech innovations, when only elite engineers with PHDs in computer science had what it takes to succeed. Back then, meaningful tech innovations weren't accessible to your everyday business or entrepreneur.

Not anymore. The modern world is driven by technology, and businesses must embrace it to innovate, differentiate, and in some cases, even survive.

Just think about it. When you last booked a taxi, did you call a company or use an app? When you last watched a movie, did you buy a tangible disk or stream it? How about Christmas presents—did you buy them in store or online, and how does that compare to what you did five or ten years ago?

There is still a market for the old ways, of course, but in many sectors, the old ways are shrinking. In this modern world, can you afford to not have a solid foundation of knowledge in tech stuff? Will you, your business, or the business you work for remain relevant in ten years if you don't adapt?

The good news is that it's no longer just elite engineers leading the way. In fact, with diversity of thought becoming ever more important to innovation,[1] it's increasingly non-tech people who identify game-changing tech opportunities.

However, the technology landscape changes rapidly and the amount of stuff you need to learn may seem vast, so I understand if you find the world of tech difficult or overwhelming. Especially how to make use of tech if you don't fully understand how it works. That's why this book exists.

Execute Your Tech Idea is for different audiences who share similar journeys — be it professionals or managers looking to innovate and grow within their existing company, founders of new start-up businesses, or those who have been around a little longer and have existing products or services. Essentially, businesses at all stages of their life journey that need help with tech ideas.

This book is called *Execute Your Tech Idea* for good reason. I want to help you to do exactly that: **execute**. We'll go on a journey from idea generation to making money, planning to marketing, and managing your product roadmap. The topics progress in maturity chapter by chapter, developing ideas as you develop your understanding.

Execute Your Tech Idea comprises three parts:

1. **Part 1: Find and Qualify Your Idea**

   Find your tech idea and identify whether it can make money or save time and money.

2. **Part 2: Implement Your Idea**

   Establish your value, prioritise your tech idea, plan the best approach, and begin your build!

3. **Part 3: Launch and Maintain Your Idea**

   Releasing your tech idea into the wild, marketing to gain users or customers, and managing your ongoing feature roadmap.

New start-ups may find themselves at the beginning of this journey, while businesses with established products or services may be somewhere in the middle. Either way, as you develop new ideas, you will repeat the cycle.

And if your tech idea involves launching a product that you wish to make money from, you'll enjoy the later chapters in Part 3 by our guest marketing expert Steve Ward. These chapters cover the basic principles of marketing and 25 marketing channels you can use to launch your tech idea.

Why this book?

I wrote this book with the 80/20 principle in mind, telling you the 20% of things that will deliver 80% of the results for you.[2] I cover the most important topics in enough depth to be effective but not overwhelming. As such, everything in this book should be easy to understand, approachable, and do-able.[3]

Most tech ideas need solid software behind the scenes, and this can take many forms including apps, widgets, tools, games, websites, web apps and portals, server software, data, device firmware, desktop or cloud software, and so on. If you're interested in any of these, then this book is definitely for you.

And if you're interested in any of these but feel a bit lost, I'll make the world of tech feel more approachable by covering the different methods you can use to build up your tech idea *and* by outlining the common terminology used in the tech industry so you don't get lost in jargon when speaking with technical teams.

I've helped hundreds of people just like you to plan and deliver successful technology projects, and this book is a culmination of a decade of experience helping those at different stages of their business journey, from single-person start-ups to senior leaders of businesses with £1bn market valuations.

Many of these businesses approached me with nothing but an idea, while others had full business plans. Some were one-man bands, others were fledgling entrepreneurs, some worked for small businesses, and some were responsible for multi-million-pound budgets. While these people seem varied, I discovered that they often face similar challenges and follow similar journeys.

In fact, start-ups need to act like established businesses as they grow, and established businesses need to act more like start-ups as they innovate. That's why the *Execute Your Tech Idea* process repeats in cycles.

Whatever stage you're currently at, I hope this book helps you feel empowered and in control, and that it helps you to achieve success on your business journey.

CHAPTER GUIDE

Execute Your Tech Idea is for professionals, managers, and start-up founders. This is because the journey for each of these people is very similar.

However, if your idea involves launching a product that you plan to sell to businesses or customers, then your motivations and interests will be different to those with internal tech ideas aiming to improve their business (not to be sold to end customers). And some ideas may sit across both.

So, to ensure that you get maximum value from this book, this list will show you which chapters relate to launching a new product and which relate to launching an internal tool. If your tech idea contains elements of both, then all chapters are relevant.

<u>Key</u>

↑   Highly Relevant

→   Some Relevance

Chapter	Relevant To	
	New Internal Tool	New Product
Part 1: Find & Qualify Your Idea Find your idea and identify whether it can make money or save time and money		
Chapter 1: Find the Winning Idea	↑	↑
Chapter 2: The Tech Supporting Your Idea	↑	↑
Chapter 3: Ways to Make Money	→	↑
Chapter 4: Is Your Tech Idea Worth Exploring Further?	↑	↑
Chapter 5: Why Your Vision Matters	→	↑
Part 2: Implement Your Idea Establish your value, prioritise your idea, plan the best approach, and begin your build!		
Chapter 6: Define Your Value Proposition	↑	↑
Chapter 7: Your Minimum Viable Product (MVP)	↑	↑
Chapter 8: Raising and Managing Money	→	↑
Chapter 9: How to Convince Key People	↑	↑
Chapter 10: Document Before You Implement	↑	↑
Chapter 11: Ways to Build Your Tech Idea	↑	↑
Chapter 12: How to Pick a Tech Team	↑	↑
Chapter 13: Project Lifecycle Types	↑	↑

Chapter	Relevant To	
	New Internal Tool	New Product
Part 3: Launch & Maintain Your Idea Releasing your idea into the wild, marketing to gain users or customers, and managing your ongoing feature roadmap		
Chapter 14: Marketing Principles	→	↑
Chapter 15: The 25 Marketing Channels	→	↑
Chapter 16: Choose Your Marketing Channels	→	↑
Chapter 17: Plan Your Product Roadmap	↑	↑
Chapter 18: Summary Checklist	↑	↑
What Next and About The Authors	↑	↑

If you're using this book as a quick-reference guide, you may choose to read only the chapters that are relevant to you. But if you're looking for the full *Execute Your Tech Idea* experience, I recommend that you read all chapters.

I like to back up what I say where possible. As such, this book contains 27 pages of over 280 end notes with links to helpful resources, links, and supporting information! So if you see a small number next to something I say that spikes your interest, then please check out the accompanying footnote at the end of the book (or touch the number on e-readers), you won't be disappointed.

PART 1:
FIND AND QUALIFY
YOUR IDEA

Find your idea and identify whether it can
make money or save time and money

CHAPTER 1:
FIND THE WINNING IDEA

"Whether real or perceived, believing you have a disadvantage forces you to find new and clever ways to compete. It's always the organizations that are resource constrained that come up with the good ideas to win."
— Simon Sinek, Author of Find Your Why

Chapter Relevance	
New Internal Tool	**New Product**
↑	↑

In this chapter, we will cover the different techniques you can use to identify a winning idea, including the different types of innovations to look out for, as well as several active and passive idea generation techniques.

Types of innovation

Most entrepreneurs and businesspeople think that an idea must be completely revolutionary for it to carry any value. It's not true. Constant incremental improvements and implementing micro-ideas are just as important, if not more so. After all, it's often easier to improve 20 things by a small amount than one thing by a huge amount. This chapter will help you to spot ways to improve what you're doing already or launch something completely new.

Radical innovation is where you completely destroy an existing business model or practice to establish your business in the new space you've created. Incremental innovation is where you make a series of small improvements to an existing business, model, or product.

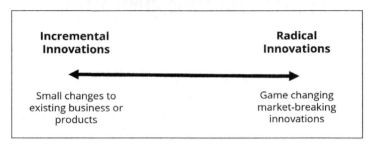

Incremental vs. Radical Innovation Continuum

Incremental Innovations	Radical Innovations
Small changes to existing business or products	Game changing market-breaking innovations

Most business innovations are not radical interventions, and you can find lots of opportunities to innovate through incremental micro-improvements across multiple areas — you don't necessarily have to have one huge game-changing idea. [4]

The tech idea you choose will sit somewhere on the spectrum of incremental to radical, and as it develops, there will be phases where it swings towards more radical or more incremental. Understanding your innovation aims will guide what to do, why, and how to search for new ideas. [5]

In 2007, roommates Joe Gebbia and Brian Chesky came up with the idea of Airbnb when they set up a simple website and rented out their three mattresses. In 2008, once they had proven the idea, they recruited former roommate Nathan Blecharczyk to join the venture and build a new version. Joe and Brian were design students with very little money and certainly weren't elite developers. Yet, they spotted the opportunity, created a plan, and formed a business that is now a $30bn global technology giant.

Is the Airbnb example incremental or radical?

Many people would rent their holiday homes to others back then, with several platforms already available to help holiday-makers find and book these locations. Airbnb grew off the back of many incremental innovations, doing a lot of small things differently over a wide range of areas to create value and drive their success. They made it easier to find rooms more cheaply, drove a unique client experience, and simplified payments and terms of engagement.

You must also be aware of which idea-generation techniques will work for you based on the type of innovation you're targeting. For example, it's often far more important to seek and use existing or prospective customer information (if you have it) when forming radical innovations rather than incremental ones. [6]

So, where can you look to find these innovation opportunities? Over 30 years of research tells us there are ten main areas: [7]

1. **Revenue model:** How and where you make money.

2. **Networks:** How and where you connect and network with others to create value.

3. **Business structure:** The structure of your business, people, and resources.

4. **Process:** Your business processes.

5. **Value proposition:** The value that your products or services bring.

6. **Ecosystem:** The way that your products or services connect.

7. **Service:** The level and quality of service that you provide around your offering.

8. **Channel:** The channel(s) you use to reach customers and how you use them.

9. **Brand:** The strength and positioning of your brand.

10. **Customer engagement:** How you engage and build relationships with customers.

You should write down and reflect on this list often, especially when you're about to embark on a process to generate and develop useful ideas. For example, you might look for ways to build a lot of passive innovations across each of the ten innovation opportunity areas where the innovations are complementary. [8]

Passive idea generation

Passive idea generation techniques are activities and best practices that you can incorporate into your day-to-day life (or business life) to increase the likelihood of spotting a useful idea.

Passive idea generation is powerful because your brain is a web of connections, and your memories and experiences exist as bundles and groups of these connections. [9] You can use the properties that emerge from this structure to help you to spot new ideas. Here is a simplified version of how this might look in your brain, consisting of many connections between brain cells (neurons):

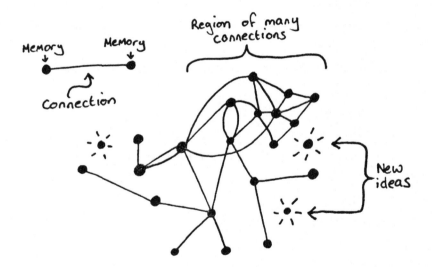

Memories and ideas get stronger when they're reinforced. Think something over and over again and you build more connections between those ideas or memories. Think about the same thing in different ways and you connect new memories together, providing different perspectives on the existing ones. This is the biology of creativity.

For example, when you read content or watch online videos, you may not actively do so to come up with new ideas, but the information learned passively allows your brain to create more varied connections. Who knows, you may only be one connection away from a major break-though. By combining the various idea generation approaches mentioned in this chapter, you increase your brain's capability to spot and capitalise on opportunities. If you implement this technique of passive learning, you might just have an "AHA!" moment whilst you're in the middle of doing something else.

Here are some techniques and good habits that aid with passive idea generation.

Passive method 1: Habitual idea spotting

Mindset and idea spotting is the process of criticising the world around you and trying to come up with interesting ways to solve issues you find. At this stage, how well the solution works isn't important—what matters is that you get into the habit of spotting issues and being creative in a wide range of circumstances. In fact, the crazier the solution the better! Flex that outside-the-box thinking muscle.

If you're struggling for inspiration, here are some funny examples from the BoredElonMusk parody account on Twitter.[10] These wacky ideas might add value if the implementation wasn't impossible or impractical:

> *"Eye-tracking blog platform that does not allow comments to be posted until the entire article is read."*

> *"Doorbell only audible to humans so your dog doesn't freak out every time a package is delivered."*

When idea-spotting, you're looking for any problems that deliver a tangible benefit to someone once solved. Get used to doing this every day in lots of different scenarios and after a couple of months, it will become a habit and you will notice them everywhere.[11]

Many of them will be useful but after a while, you may spot a real golden opportunity.

Passive method 2: Step out of your comfort zone

To further double-down on the idea-spotting technique, try deliberately placing yourself in situations outside your comfort zone. It's human nature to reduce, mitigate, or outright avoid risk, which results in us forming habits and engaging in familiar activities, often on autopilot. However, this approach of avoiding risk and uncertainty is counter-intuitive, as research shows that this behaviour seriously limits creativity, and you want maximum creativity when idea-spotting! [12]

Avoidance behaviours prevent you from seeking new ideas and stop you from seeing otherwise obvious new opportunities. The most successful businesses in the world would not have got where they are today without embracing uncertainty and taking measured risks.

We may think that risk avoidance behaviour limits our risk, but it may do the opposite. In business and in life, it's impossible to completely avoid all risks. [13] Of course, individuals and businesses have different levels of risk tolerance. However, even for a service business owner with low startup costs and monthly outgoings, they must quit their job and sacrifice a stable income as their venture grows and demands more time to support its success. This is a significant commitment given that many businesses and business ideas don't make a penny in profits until years after they start up.

Logistics giant Amazon is a great example of strategic risk-taking and the risk of not taking risks. They were founded in 1994 but didn't make a profit for over 20 years. Now, they have a culture of risk-taking and constant reinvestment. Founder and ex-CEO Jeff Bezos famously said:

> *"If you only do things where you know the answer in advance, your company goes away."*

Exceptional long-term results requires short-term uncertainty. So, explore ideas that risk financial loss, follow thought patterns that feel unfamiliar, and do things that put you outside your comfort zone.

If you're struggling to think of ways to step out of your comfort zone and embrace uncertainty, then first think of the outcomes you'd like to see and work backwards from there. What is something you'd like to see happen? Now close your eyes and really visualise it. Walk backwards from that vision and imagine the steps needed to get there.

For example, if you want to be self-made millionaire in the food and hospitality space, you could start with the vision of what your business looks like, what it's doing, and your daily agenda. Next, imagine the step before that vision such as the products and services you created to get there (and hopefully the technology to get you there too!).

As you work backwards, you'll identify things that you hadn't considered, challenges you need to overcome, and times where there are many options available but each carries a different risk.

Stepping out of your comfort zone doesn't mean taking silly risks, and I certainly don't want to see you take a risky gamble. You should be looking for sensible measured risks — the ones where even if you fail, you at least walk away a better, more experienced version of yourself than when you started.

Passive method 3: Build your networks

Going it alone can be tempting, especially if you're an introverted person by nature. However, second opinions are extremely valuable, and discussing your business idea with people is a great way to build relationships if approached the right way. The bridges you build with people early on in your business, idea, or career can stick with you for a long time, adding value far into the future in ways that you can't predict.

So, you should build your networks for reasons other than idea generation, but through networking, you will inevitably spot ideas and opportunities. That's why this is a passive method.

One of the easiest ways to start the networking ball rolling is to arrange an in-person one-to-one meeting over coffee. This can include friends, colleagues, family, friends of friends, colleagues of colleagues, neighbours, business owners near you, or even people you bump into at the local coffee shop.

If you're extroverted, then striking up a conversation with everyone and anyone may come naturally to you and feel like fun. If you're introverted to the point that even thinking about networking fills you with dread, here are some simple low-risk ways to get started and eliminate that fear and dread bit by bit:

1. **Make LinkedIn connections** [14]

   LinkedIn is a social network aimed exclusively at business connections. Sign up for an account if you don't already have one. Invite people you know or have worked with to connect with you. When someone accepts your connection request, send them a short message thanking them and ask if they'd like a catch-up over coffee. Alternatively, you can start by messaging the people you are already connected with.

   Some people use LinkedIn to arrange meetings to sell what they're offering, which might put your contact on guard, so make it clear that the person won't be sold to and that you'd like to hear about what they're up to.

2. **Ask for introductions**

   If the average person has 200 friends on social media, [15] and those friends each have 200 more friends. This means there are approximately 40,000 people that you are directly connected to or could be introduced to by one of your direct connections—that's a lot of people! [16]

   I'm willing to bet that at least a handful of them will be interested in what you're doing and in speaking with you about it. Say something like *"Hey Zoe, I saw that your friend Keith works in the Automotive industry. I'm launching a new project which automates quality checking of car parts and would love to ask them a few questions about what they've learned operating in this sector. Do you think you'd be able to introduce me in a message please?"*

   However, 40,000 is a lot of people, don't start messaging people at random. You should focus your energy on trying to contact and make connections with people most closely aligned with what you're trying to achieve. Maybe they're involved in similar projects, are connected to people you know, or have customers in a similar industry to yours? If you're struggling with this, then our later chapter "Chapter 6: Define Your Value Proposition" will guide you.

3. **Attend a local networking event**

Most cities and small towns have a ton of networking events you can attend. In fact, there are often well-established networking groups already in place that structure the process of networking. [17]

See if you can get hold of a guest list beforehand and identify a shortlist of three to five people you'd like to speak to. At the event, speak to the people on your shortlist, find out a bit about them, and ask if they'd attend a one-to-one meeting. Most of the popular networking groups encourage visitors to have one-to-one meetings, so it shouldn't be difficult to find a few people who agree to meet you.

4. **Attend an online event**

I don't know if you've noticed but not too long ago, the world experienced a major pandemic, which changed a lot of people's attitudes towards events conducted online via video. Today, there are thousands of online events you can attend across a huge range of topics, and you can choose to attend them live and communicate with other attendees or watch pre-recordings online.

If you pick the right kind of event, they will often focus on issues and opportunities happening in that space right now, and you can gain insights otherwise inaccessible via books and the media. [18] Attend at least one event in your category of interest and see if you can spot any problem statements (short sentences that draw your attention to a problem that exists in the market). One of these problems could become your next opportunity.

Whilst networking, you will find yourself exposed to a wide variety of people with very different capabilities and experiences to yours. It's all part of the game. Take the time to listen to their challenges as every pain point is an opportunity to innovate. Likewise, focus on spotting new technologies as you may be able to build on these ideas to create something of value or apply the lessons across different industries.

However, networking isn't just about speaking to as many faces as possible at a single event within a large room. It's about building relationships with interesting people and taking the time to understand what each person is trying to achieve. Large events should be a platform to selectively arrange strategic one-to-one meetings for mutual benefit, but large events aren't always necessary. Quite often, the best results come from speaking with people you already know or people they know.

In fact, gathering people around you is essentially a form of networking, and it's valuable beyond just generating ideas. If done correctly, it can positively impact your life in unpredictable ways. Think about how many events have happened in your life because you met someone by chance.

For example, a few years ago, my friend was struggling to find a place to live and was keen to explore people in my networks, so I asked my landlord friend if he had a spare room in any of his properties, which he did. While there, my friend met someone who was also renting a room in that house, and they recently got married. Without this impromptu act of networking, they would not be married today!

> "You can design and create and build the most wonderful place in the world. But it takes people to make the dream a reality." – Walt Disney

Ultimately, the best way to build long-standing relationships is to be selfless in your approach to networking. Respect that other people's time is valuable and try to identify ways you can help them in return for their time. [19]

If you start gradually building a network of people around you as you launch your idea, then it will be vastly easier to get feedback as the idea develops. If you're lucky, then some of these connections might even become users or customers.

Passive method 4: Engage in ongoing personal development

If you widen your skills, knowledge, and experience, then you'll increase your ability to spot ideas in those areas. The more deeply you understand a topic, the easier you can solve problems related to it.

Imagine you want to put up a shelf. If you're new to DIY, you might know that putting a shelf up is supposedly easy but have no idea about the various methods of attaching it to the wall, something that even the most novice builder would know. Did you know there are different kinds of screws and wall plugs, some that only go into plaster and others that anchor all the way into the brick, some that can cope with lightweight items and some heavy? Putting up a shelf to hold a small plant pot requires a different approach to one that needs to hold something weighty.

Even for something as simple as putting up a shelf, a relatively small amount of initial learning makes a big difference to the way you complete the task. Imagine what growing your knowledge in business and technology could help you to achieve?

So, before you jump in head-first and try to identify opportunities for innovation, make sure you have a basic understanding of the area you wish to target. It's better to research the different kinds of shelving and wall plugs before installing that shelf, as it could prevent you from accidentally tearing down a wall.

There is so much information out there today that it doesn't take long to learn the basics in any area. Take some time to explore the technologies, tools, and software that exist in your sector. See how others have used them, find out what is important to people in that industry, and see whether there are any books or videos available that cover best practices.

For example, in writing this book, I discovered a wealth of videos from people who had launched books before, covering the best practices of book structure, writing approaches, and authoring tools. A few afternoons watching videos on the topic made a huge difference, enabling me to avoid common mistakes by starting from the strong foundations of best practice identified by others.

So, well done in taking the time to read this book! You clearly already understand the importance of learning the foundations before you get started.

Here are some useful sources of passive learning that you can use to improve your passive idea generation abilities:

1. Read quick-start and guide books on the industries you're interested in.
2. Subscribe to YouTube channels and search for how-to and explainer content.
3. Join an on-topic social media group or forum.
4. Take a short online course. [20]
5. Search for industry papers and e-guides (both online and offline).
6. Subscribe to a think tank publication and read their content regularly. [21]
7. Ask an experienced colleague or connection if they will mentor you.
8. Combine personal development with your networking activity and arrange one-to-ones with people in your target area for their experience and guidance.
9. Sign up for a Reddit.com account, follow subreddits that align with your business interests (avoiding social-media-like meme content), and read it regularly.
10. Search for related podcasts and subscribe to them.
11. Read online wikis and follow reference links in each article to expand your knowledge.

With personal development, consistency is key. It's something you need to do all the time and build into your daily habits; that's what makes it effortless.

It would feel daunting if I asked you to be doubly as good at something you're currently doing. Why? Because I've given you a mountain to climb with no path to get there. But if I asked you to improve by just 1%, that suddenly feels a lot more achievable. Why? Because it is more achievable! If you can make yourself just 1% better today than yesterday, you'll double your effectiveness in 70 days, which is just over two months. [22]

Aim to do a small amount often rather than a huge amount all at once and find ways to build this into your routine. Maybe you can get into the habit of listening to industry podcasts on the way into work or swap the time you'd normally spend watching the news with relevant, informative YouTube videos instead.

Active idea generation

Active idea generation involves completing tasks and activities with the objective of coming up with ideas immediately in the moment. This active approach to idea generation is more methodical than the passive methods, relying less on luck and in-the-moment inspiration and more on considered thought.

Active method 1: Idea audits

Brainstorming is an idea auditing technique where you connect concepts to come up with new ideas, usually by exploring things associated with the problem you're trying to solve. For example, if your problem is how to come up with the perfect tech idea, you could begin by listing all of the possible approaches to come up with that idea. This encourages you to think creatively and coming up with ideas you might not have thought of before.

It's common to draw a brainstorm diagram on paper as part of this process, like this:

Example Brainstorm / Mind Map

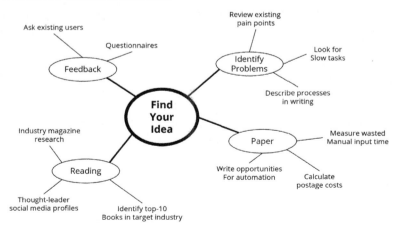

Brainstorming is a creative process, and research shows that those who use it as a method of idea creation come up with more ideas than those who don't. [23] In addition, those who brainstorm in groups identify significantly more ideas than lone brainstormers.

Here are some best practices to get the most out of your brainstorming session:

1. **Schedule it**

   Set a date and time to conduct the session, and allow for about an hour.

2. **Gather a team or group of collaborators to help you**

   Get people together who can contribute to the brainstorming session. Around 5-7 people is ideal, but the process can work with fewer. Try to get people involved who have a mixture of personality types, backgrounds, and skills. If you don't have an existing team, then you could pull together friends and family, and maybe bribe them with pizza. Or you can create a focus group where you pay people to help you brainstorm. This is slightly different to collecting feedback as the group should be more actively involved in helping you shape your high-level ideas and concepts.

3. **Facilitate and educate**

   Brainstorming sessions with a facilitator can generate more ideas. Make sure everyone knows the process and what you're trying to achieve. Let them know you will be leading the process. If you can, write down the problem you're trying to solve. If you haven't identified a problem yet, your brainstorming session can focus on finding the problem.

4. **Be stupid**

   No ideas are stupid ideas, so keep an open mind and push boundaries. Criticism is not allowed. Prepare some questions in advance that encourage people to identify varied ideas, for example, "What is our Ten-year plan, and what is stopping us achieving it in 6 months?" [24]

5. **Aim for quantity**

   Set a target of how many ideas you hope to generate, and be ambitious. You want people to have a goal to work with that encourages them to think quickly.

6. **Evaluate**

   Research shows that sleeping on a complex problem increases your effectiveness at solving that problem, so schedule time to review the session notes the day after. [25]

Active method 2: Value audits

Value audits are where you ask questions about the value that something brings in order to identify new opportunities. You can value audit the concepts you identify in the early brainstorming session, as well as problems you've identified, or solutions you've spotted. Consider the following questions:

1. **Can it prevent the risk of loss (or perceived risk of loss)?**

   Loss aversion is a powerful force, and behavioural economists have discovered that people will take more risks to avoid a loss than they would for the chance to make a similar-sized gain. [26] Also, people place more value on the things they already own compared to identical things that they could own. [27] So, provide significant value by creating a product or service that helps avoid the risk of loss.

2. **Can it offer a greater feeling of control?**

   People like feeling that they are in control as it's rewarding and removes anxiety. Can you identify ways to give someone greater control without it being overwhelming? Sometimes, the act of giving a user a suitable level of control requires hundreds of features and options, especially if they're a power-user. But depending on the demographics and capability of the user, you may be able to give them a feeling of control by presenting a smaller number of curated options. Giving the user too much control over a process often means more complexity and a higher cost to build your solution. Find the right balance and understand what your target audience values; that way, you can offer enough control to keep them happy but not so much that it's costly to implement.

3. Can it save someone time?

Every hour spent doing something that can be avoided could have been spent doing something more productive. This is called opportunity cost. Can you think of ways to reduce the amount of unnecessary time someone spends on a task?

4. Can it give someone confidence over the competition?

A product's brand is its promise to the customer. I know that if I buy my favourite designer pair of shoes, they will fit, be of premium quality, and not wear out in six months. I know they're slightly more expensive but I'm happy to pay the extra to enjoy the value that the brand promises. If I bought a cheap pair, they might break in a few months and mean I have to buy another pair, or the fit may be poor. Are there ways you can deliver confidence that others can't?

5. Can it improve someone's status?

Achieving a higher personal status is a fundamental human motivation, and many people prefer to have a higher status than those around them, even if it's caused by those around them having a lower status. [28] People therefore like purchasing things that elevate their status. Facebook (now called Meta) was founded on this principle, as it was originally an in-club exclusive to Harvard Students, which propelled the platform's early popularity. Once it gained traction, it held its dominant position due to the network effects of social media. [29]

6. Can it save money?

This might be an obvious one, but if you can create a product or service that costs £X but saves £XX, then the value it brings is the difference between the cost and the money saved. If the difference is large enough, then people would be silly not to buy or implement what you have to offer.

7. **Can it enable something to happen that can't currently happen?**

   Enablement carries value. At one-point, GPS technology didn't exist in the consumer market, but someone invented it and then created the satnav for cars. This ground-breaking technology provided instant step-by-step and map navigation to drivers who previously had to rely on a giant map that lived permanently on the back seat.

8. **Can it improve performance?**

   For example, can you enable someone to do their job better? Can you make their car accelerate faster? Can you get results more quickly? If the performance you improve has value to the end user, then your idea will have value relative to that performance improvement.

9. **Can it add moral value or help the user to help others?**

   If your product or service aligns with a person's moral values, you can become part of a captive market for that kind of person. For example, a vegan isn't going to buy a sirloin steak from you no matter the price or the quality, but make an animal-free alternative and they may buy nothing else. Can you make your problem greener, fairer, or aligned with another socio-economic issue? [30] Or can you align with a different value system, such as political values?

10. **Can it deliver something more cost-effective than the alternatives?**

    In my opinion, being the cheapest is one of the worst ways to differentiate yourself from the competition. However, adding as much value as possible for a competitive price is different, as it maximises value for the consumer. Consider whether there are ways you can add more value to the customer at a competitive price. It doesn't have to be cheap, but if the end user feels that you offer the best value for money, then you have a business.

11. **Can it make something more accessible than the alternatives or better align with the user's habits?**

Imagine you have two competing types of account software that perform the exact same functions. Which one are you more likely to use: the complicated one or the simple one? In the UK Finance and Technology sector (Fintech for short), many of the current challenger banks have established their market share on the principle of simplicity. [31] They implemented processes using apps that enable clients to sign up and access their accounts quickly, without the need to visit a branch.

12. **Can it offer something fun or entertaining, or deliver experiential value?**

Entertainment isn't just about fun — it's also about the elimination of boredom. People would rather experience pain than boredom, with 67% of men and 25% of women in one study opting to receive an electric shock rather than sitting in quiet contemplation for 15 minutes. It's no surprise that people pay big bucks for entertainment in various forms. [32] For example, in 2020, video streaming service Netflix gained 15 million new subscribers in the first 3 months of the year during the pandemic, doubling previous Wall Street estimations. [33] Make the world stay at home and people will buy things to alleviate the boredom. Social media, dating, gaming, video streaming, and news sites are all examples of tech that helps users have fun or eliminate boredom.

13. **Can it fulfil someone's desires or give someone something they want?**

If people want something and are driven enough, they will be prepared to allocate resources, time, or money, to get it. Take a look at the 20 top-grossing apps at the moment in your smartphone app store. There's a reason why so many of them are dating apps. To carve out unique value, consider ways to align your idea with a fundamental human desire and help people meet that they desire.

14. **Can it aid someone's training or personal development?**

Some innovations help your customers, staff, or end users to grow. For example, online learning systems that aim to speed up how quickly businesspeople learn and retain information. If people can become effective at their jobs more quickly, this gives businesses more time as the person is more productive, which is worth real money. Can you enable someone to pick up a skill more quickly, better consume new information, or gain vital experience?

15. Can it give the user insights to improve their decision-making?

When you create software, you collect data, which raises opportunities to display that data in ways that drive user behaviour. For example, I was at a local tech event where a major UK rail network talked about how they'd installed sensors on various parts of the carriage to record events and associated data. They could have presented this raw data to the user in a table form, with a list of all events, but would their users get lost in all the detail? Instead, they developed insights by looking at trends in the data to alert users about when to act. For example, they discovered that if a door took longer to close than normal, it was likely to fail and require maintenance, something that could put the carriage out of action for days and cost the rail operator a lot of money. By identifying this correlation, they could take pre-emptive action. Could you collect data and identify metrics to provide similar benefits to your organisation or clients?

16. Can it reach people differently via marketing compared to the incumbent?

In a 2015 "peep show" experiment, YouTubers Hamish & Andy teamed up to offer a one-on-one music session with world-famous artist Ed Sheeran. The end result was disappointing to say the least but proved a very interesting point...

Scan this QR or visit the URL to watch the video:

Watch The Video

See how Hamish, Andy, and Ed Sheeran help demonstrate my lesson

executeyourtechidea.com/r/edvideo

If you don't have the time to watch the video, Andy set up a genuine once-in-a-lifetime experience for people. He had the real-life Ed Sheeran in a booth in a shop behind him with the curtain closed. He approached people walking past on a busy street and asked if they'd like to pay $2 to see Ed Sheeran perform for them. Despite this offer being incredible value for money, everyone who Andy asked turned down the offer. As you can imagine, some super-fans would probably be willing to part with thousands for the privilege!

The lesson of the story is that you could have the absolute best product in the world and nobody will buy it if your marketing and sales messaging is poor. Andy's approach to sales and marketing was poor, and therefore nobody took him up on his offer despite its huge value for money.

Similarly, mediocre products can achieve great commercial success over others simply because of their approach to marketing. There's a reason why so many celebrities develop their own branded fragrance — it's very easy for them to reach and sell to their existing audience, even though the product they're selling is nearly identical to thousands of others on the market.

Your unique ability to reach a market can drive value. Can your idea leverage this to gain traction in ways unattainable to your competition?

Hopefully you can say yes to one or more of these factors. Whilst answering, you should also consider the ten types of innovation we discussed earlier as the benefit that your solution brings will fit into one of those ten categories.

Let's apply this value audit to a real tech innovation that has been around for a long time now — WhatsApp. It's an instant messaging app that sends messages over a user's data plan rather than traditional SMS or MMS messages. On first inspection, its value isn't any different to an SMS message, in fact, it could be perceived as worse as you can only message users who also have WhatsApp. Then why was it so successful? [34]

1. **It saves money:** There is no cost, other than data, to send picture and video messages, something that phone networks typically charge for.

2. **It appeals to human nature:** There is in-built virality in WhatsApp. If one person likes it, they're likely to share the app so they can message their friends. When someone receives an invite to use WhatsApp, it appeals to their human nature, driven by social needs or status — especially if their friend has sent them a message they can't read otherwise.

3. **It enables actions otherwise not possible:** SMS and MMS messages don't allow people to send high-quality video while WhatsApp does. It also allows users to create group messages, which is not possible using older technologies.

If you can spot a problem and identify the value in solving it, then you can begin to come up with ideas around *how* to solve it.

Active method 3: Time audits

Saving time is such a powerful driver of value that it deserves its own audit methods.

One of the best places to look for opportunities to save time is the things you do every day. Take time to review your daily tasks and identify any that are manual or repetitive. If you're doing something that involves a lot of 'copy and paste', then you may have spotted an opportunity to innovate.

In business, it's common to create spreadsheets to organise your business processes. Even businesses that use lots of different tools and software will plug holes in what these tools provide by creating a spreadsheet. If you aren't careful, you risk creating more problems than you solve, left with the overhead of manually synchronising information between the two — copy and paste, copy and paste, copy and paste... There can be a lot of waste in copy and paste.

To put this into perspective, a computer is much faster at copying and pasting than you are, so if you have the option to automate it, why wouldn't you? Even 10 minutes lost to automatable tasks per day equates to 40 hours over the course of a year. A team of ten people wasting 30 minutes each day are losing nearly 40 working weeks. I wonder what that may be worth to your business? Always automate before you delegate.

Self-service checkouts at supermarkets solve a similar problem — they take a repetitive task and get a computer to do it. Technology has come so far that even cars are capable of automatically driving. If problems this complex can be solved with automation, you can be confident that there are probably automated solutions to your problems too. It's usually not a question of possibility — it's a question of money.

Active method 4: Process audits

Process audits are similar to time audits but where you audit other kinds of waste, such as lost money, resources, or unnecessary complexity.

Let me give you can example that blew my mind: I once had a prospective new customer, who we will call John. John and I were talking about starting a new project together, and over the course of a couple of weeks, we'd sent a lot of emails to clarify what was needed. I visited John in person to clarify some of the finer details and get the project started. Everything was going well until I was accused of not responding to an email containing an important attachment.

I must have missed it; we've all been there. So I apologised and asked for more details on what was sent. Rather than search for the email on his phone or laptop, John responded by reaching under his desk to retrieve a ring binder, full to the brim with every single email and attachment ever sent between us printed and slotted in little plastic pockets.

Him printing every email he'd ever been sent was so wasteful and unnecessary that I'm scarred for life — I still have nightmares about it.

Paper forms and processes are another evil of this world. You write it, print it, put a stamp on it, post it. A post office receives it, sorts it, puts it on a van, and a person manually posts it. The other person gets it, reads it, hand-writes their details, puts it in an envelope, stamps it, sends it back. The post office receives, sorts, transports, and posts again. You receive it, sort it, scan it, or manually type it in, probably making some errors along the way.

That paragraph was a nightmare to read, and it's even worse to go through the process. It's horrible. Kill it with fire. And don't get me started on the carbon footprint!

The digital alternative? You email it. They complete it online. Done.

So, look for ways that you can spot wasted capital in a process, whether that's money, resources, people, or anything that is a resource to your business or prospective customers.

Active method 5: Business audits

There are many different business tools and frameworks that you can use to reflect on what you're doing or what you plan to do and the SWOT analysis is probably one of the most powerful techniques to help you form a strategy. [35]

SWOT stands for **S**trengths, **W**eaknesses, **O**pportunities, and **T**hreats.

It's a quick way to gain an overview of the competitive environment that your business operates in. When you create a SWOT, you identify factors under each of these four sections and it forces you to think objectively about what you're currently doing.

A SWOT analysis is a great way to help you generate new ideas, prioritise, identify barriers to your goals, and plan a strategy to overcome those barriers. If you have a few ideas you're trying to decide between, a SWOT diagram can be a great way to help you decide which idea you want to go with.

If you don't have an existing business, then I recommend creating a SWOT based on a competitor in the space where you wish to operate. If you have an existing business or your idea is based on replacing an internal process, then start by completing a SWOT for that process. By looking at the existing competitive environment, you will identify opportunities for innovation that influence your ideas. You can then decide how to approach your own idea and complete a SWOT analysis for that too.

Get started by drawing a two-by-two grid with each letter of SWOT: Strengths, Weaknesses, Opportunities, and Threats as headings in each section of the grid:

Strengths	Weaknesses
Opportunities	Threats

Once you've drawn your grid, you should populate each box, taking care to think about your business, idea, and the competitive environment that you operate within.

Let's take Netlifx as an example. They have an app for most platforms, including a web player. They grew extremely popular by being one of the first providers to give users a vast range of shows to watch, breaking away from the traditional model of watching shows at a set time on a television channel or grabbing a movie from your local video store, an activity that is now a relic of the past.

Let's go at a time and build our SWOT analysis based on a young Netflix, a time before the physical video rental chain Blockbuster went bust. [36]

Strengths

- Gives users immediate access to films and TV content on demand.

- No need to go to a brick-and-mortar shop.

- No late return fees for forgetting to return your movie.

- Thousands of titles available.

- Ability to set up a global presence without the infrastructure costs of leasing or building thousands of stores.

- Pay-monthly billing model generates a reliable source of recurring revenue.

Weaknesses

- Hasn't got an established brand yet.

- Not as much cash available to fund growth objectives compared to larger, more-established businesses operating in the marketplace.

- Significant technology infrastructure is required to scale globally.

- Acquiring licences from content producers may be costly, especially global ones, which could be a challenge when monthly fees are relatively low.

- Requires users to have a high-speed internet connection.

- The quality of online video is lower than on BluRay discs, especially over a slow internet connection.

Opportunities

- Become known as the fastest-growing and most popular online streaming service.

- Acquire licences for lots of older content relatively cheaply as content producers seek to monetise their historic movie libraries in ways they cannot at present.

- Source a handful of major movies that can only be found on Netflix, giving users a reason to try the service.

- Raise money via investment (e.g. private equity) to enable aggressive investment in video-streaming infrastructure.

- Develop exclusive content that can only be found on Netflix, giving users who have already subscribed a reason to remain a customer.

- Develop apps for games consoles, allowing the app to be available in the living rooms of anyone with a PlayStation or Xbox.

Threats

- A major competitor with more money could enter the market faster.

- Other large competitors could begin developing exclusives that are only available on their platform. This could cause users to change which service provider they use.

- Abuse of the service could result in costly legal challenges due to licence disputes.

- Content licence costs could work out prohibitively expensive, clashing with the pay-per-month business model.

- May not be able to agree licence deals for enough content, so users might not see enough value in the service once they've watched most of the video library.

- Mobile phone networks could block Netflix traffic as video content consumes a high amount of internet bandwidth.

- The major games consoles may block a Netflix app on their platform in favour of their own proprietary apps, making it difficult and expensive to get the Netflix streaming app into the average living room.

Here is what this SWOT analysis looks like when plotted on our two-by-two grid, with each point simplified to be easier to read at a glance:

SWOT Analysis Example (A Young Netflix)

Strengths	Weaknesses
1. Gives users immediate access to films and TV content on demand 2. No need to go to a brick-and-mortar shop 3. No late return fees for forgetting to return your movie 4. Thousands of titles available 5. Ability to set up a global presence without the infrastructure costs 6. Pay-monthly billing model = recurring revenue	1. Hasn't got an established brand yet 2. Not as much cash available to fund growth objectives 3. Significant technology infrastructure is required to scale globally 4. Acquiring licences 5. Requires users to have a high-speed internet connection 6. The quality of online video is lower than on BluRay discs
Opportunities	**Threats**
1. Become known as the fastest-growing and most popular online streaming service 2. Acquire licences for lots of older content relatively cheaply 3. Source a handful of major movies that can only be found on Netflix 4. Raise money via investment (e.g. private equity) 5. Develop exclusive content that can only be found on Netflix 6. Develop apps for games consoles	1. A major competitor with more money could enter the market faster 2. Other large competitors could begin developing exclusives that are only available on their platform 3. Abuse of the service could result in costly legal challenges 4. Content licence costs could work out prohibitively expensive 5. May not be able to agree licence deals for enough content 6. Mobile phone networks could block Netflix traffic 7. The major games consoles may block a Netflix app on their platform in favour of their own proprietary apps

You can download a quick-start SWOT analysis template for free from the Execute Your Tech Idea website.

Free Resource Pack ⬇

The Execute Your Tech Idea website contains further information, quick-start document templates, and other helpful free resources

executeyourtechidea.com

Once you've completed your SWOT analysis, you can explore what you've identified. See if it's possible to further improve your strengths, address your weaknesses, exploit the opportunities you've identified, or implement mitigating measures to eliminate or reduce the risk of threats.

For example, an opportunity might be to reach a wide audience by targeting a global market. Your tech idea could enable this kind of global strategy as apps and websites are accessible everywhere in the world. Similarly, if a competitor in your space is a threat to your business because they're cheaper, you may want to implement technologies to give your customers, potential customers, or users the best possible experience so you can justify charging a higher price for the premium service.

This ties back to what I said earlier: "You don't find the best ideas by trying to find the best ideas, you find them by searching for the most pressing problems then solving those problems."

Active method 5: Customer audits

Customer audits are any method where you involve the customer or prospective customer in the process of coming up with ideas. This can include identifying problems, coming up with solutions, or testing your plans or technology. Think of customer audits as a way to get targeted feedback.

Seeking feedback on a broad topic can penetrate every stage of your journey so take note — its usefulness is not limited to early idea generation. In the context of active idea generation, you should find opportunities to gather feedback about the things you're doing already.

For example, you may ask clients for feedback on an existing product or service, which might become a spark to influence your ideas, enabling you to address a problem or opportunity that presents itself in the feedback.

"Software innovation, like almost every other kind of innovation, requires the ability to collaborate and share ideas with other people, and to sit down and talk with customers and get their feedback and understand their needs."
– Bill Gates, former CEO and founder of Microsoft.

At the idea generation phase, you should be looking for feedback on existing markets, products, services, and processes. Feedback amounts to people's perception of what they experience. You don't need to have a product or an existing business to start receiving real-world customer feedback.

There are plenty of products and marketplaces that people are complaining about. Every complaint is an opportunity for improvement, and every "it won't work because..." is an opportunity for innovation. In solving these objections, you differentiate your project and create 'barriers to entry' that make it more difficult for your competitors.

"It's important to actively seek out, and listen to negative feedback, particularly from friends. Positive feedback should be water off of a duck's back, really under-weight that, and over-weight the negative feedback. Take the approach that you are wrong. Your goal is to be less wrong."
– Elon Musk [37]

There are many ways to collect feedback, and here are a few I've used over the years:

1. **Ask friends, family, and colleagues for "ugly baby feedback".**

   Nobody is going to tell you that your baby is ugly unless you explicitly ask them to do so. Ask your friends and family to provide honest, unfiltered feedback, and listen to it. Let them know it's ok to tell you if they think you have an ugly baby. Take note of what they say and look for solutions to their objections.

2. **Online surveys and questionnaires.**

   There are many websites and apps that enable you to create an online survey for free and send it to friends and colleagues. Ask a mix of questions that allow you to build up and compare scores (quantitative) and are open-ended to encourage a more personal response (qualitative). [38]

If you have an existing business, you could create a survey and build into your business processes that it should be sent to customers, prospective customers, or staff at a certain point, making feedback generation automatic.

3. **Ask your customers or prospective customers.**

Did you know that 43% of people don't leave feedback because they don't think the business cares? However, 81% would give feedback if they knew they'd get a quick response. So, people are willing to help, but be honest and keep it personal.

You can ask customers and potential users for feedback over email, phone, or in person. Understand why you are asking, and only ask questions that you intend to use. A small amount of open-ended or leading questions is best, as it gives the respondent the opportunity to provide more detail should they want to.

4. **Social media**

Welcome or not, the internet is full of people sharing their opinions publicly. Join a Reddit subreddit, LinkedIn, or Facebook group that is aligned with your market, or search Twitter for hashtags that relate to your target demographic. [39] Twitter has particularly useful advanced search tools, so you can find what people are talking about by hashtag (topic), date, location, and more. [40] Find people in your demographic, see what they're talking about, and engage them in conversation.

5. **Metrics, analytics, and usability testing**

Digital things often collect data. It's very easy for a technical person to install metric and analytic recording tools onto websites, software, or apps. Reports can tell you about things that users won't. For example, you may notice that visitors to your website aren't staying very long or that part of your system isn't working properly. Make sure you have some way to collect and review usage data.

6. **Automatic feedback as part of your processes**

If you already have people interacting with you in some way, then look for the points where you communicate with them. For a café, this could be when you bring over the bill. For an online shop, it may be an order confirmation email. Use these triggers as an opportunity to ask for feedback.

7. **Other websites**

Seek inspiration from sources such as online forums, product reviews on e-commerce sites such as Amazon or app stores, or comments on competitor's blog posts. You can set up automatic alerts in minutes and receive an automatic notification when someone creates a web page or talks about you or a topic you're interested in online. [41]

8. **Focus groups**

Bring together a group of people who are willing to listen and provide feedback in a relaxed environment. You do need to offer some incentive to encourage attendees, though it doesn't always have to be a financial reward. You'll find it easier to recruit people if you offer to pay for their time and expenses.

Focus groups have a reputation for being expensive or difficult to organise, but these days you can hold them cheaply online using free chat and video tools that allow you to invite multiple people. [42]

It's good to get a balanced range of perspectives, so I recommend exploring a range of sources to find people for focus groups, including: existing customers, prospective customers, target users, social media, and low-cost freelance websites. The objective is to create an environment that promotes opinion sharing.

9. **Staff one-to-ones, and team meetings.**

Effective managers have one-to-one meetings with their team on a weekly basis. [43] If you manage a team, then add seeking feedback into your agenda or suggest it to your superior. Staff that you line manage and that report into you (Direct reports) should be given half the meeting to discuss their own agenda, ideally from notes they've prepared themselves. In giving them autonomy to talk about whatever they please, you'll get all kinds of feedback and suggestions that would not have been raised in other settings.

Once you've collected all of the feedback, sift through it for objections and list them. If you have a lot of feedback, then tally the number of times an objection has been repeated. You should quickly start to get an idea about where the opportunities are. Getting feedback is iterative, so don't be afraid to gather more once you've made changes.

I want to emphasise this again: the barriers to entry and the objections you face are desirable — provided you can overcome them. New innovations exist because entrepreneurs come up with creative solutions. Embrace these roadblocks, because if you can learn to jump over them, then your competition might not.

Closing words

Now you are armed with different techniques to come up with ideas, set aside time to do the active techniques and try to get into the habit of following the passive techniques every day. You should regularly perform these techniques forever, because even when you've found your tech idea, your plans, markets, and target customers will change, and you will be able to spot and move with these changes.

If you constantly seek innovation, you will continue to find both radical and incremental opportunities to innovate, which will bit by bit increase your chances of success.

Key takeaways:

1. Tech innovations can be radical, incremental, or somewhere in between. Some methods of idea generation work better for radical innovations than incremental ones (and vice-versa).

2. There are both passive and active ways to generate ideas, and you should try to use both methods as appropriate.

3. The types of passive idea include mindset, processes with your team (or people you know), and ongoing personal development. You can do these things to get better at spotting and evaluating ideas.

4. Active idea generation techniques include brainstorming, reflection, process reviews, business analysis (such as SWOT), and searching for client or market feedback.

5. Embrace roadblocks and difficulties and achieve success by embracing overcoming them in ways others can't.

CHAPTER 2:
THE TECH SUPPORTING YOUR IDEA

"We have a hunger of the mind which asks for knowledge of all around us, and the more we gain, the more is our desire; the more we see, the more we are capable of seeing."
— Maria Mitchell, Astronomer

Chapter Relevance	
New Internal Tool	**New Product**
↑	↑

Understanding the different technologies that exist and how they fit together is a useful tool to help you to identify or improve your idea. This knowledge gives you the tools to creatively solve the business problems you identify.

The goal of this chapter is to help you to understand the absolute basics of what each technology is capable of, so that you can apply it to your idea. And remember, you don't need to be a tech expert to manage a tech project.

The more you understand the various technologies and platforms that exist, the better you will become at spotting opportunities to use emerging technologies to solve problems within your organisation.

You will hear the term Technology Stack, or "Tech Stack" a lot, and it means the combination of technologies that you use to deliver your project. The Tech Stack is made up of frameworks, programming languages, and tools. Think of it like the ingredients required to make a meal and the tools used to cook it. A Cook Stack might need an oven, a frying pan, rice, chicken, vegetables, and herbs. Your Tech Stack is the technical ingredients required for your project.

Technology types

Here are some simple types of technology to start you off, including what each technology or platform is and how/why you might want to use it. Chances are that your tech idea will require one or more of these types to work properly, so listen closely!

Systems

System is a broad word in the tech space covering any combination of technologies that performs a repeatable task. Apps are systems, as are websites, software, etc. Even manual processes can be a system if there are rules on what should happen and when.

In business, it's good practice to think in systems, and the most successful businesses grow their revenue per head by consistently looking for opportunities to create systems that repeatedly deliver what they do to a consistent standard (whether manual using people or automated using software).

Software

Software is any form of code that runs on a computer to perform a task. The operating system on your computer is built using software, as are the apps on your phone and computer, including your word processor or spreadsheet tools. Likewise, any website you log in to if you want perform a task is software... I could go on. If you have a tech idea, it will most likely involve creating or making use of software.

Online portals & progressive web apps

Online portals, also known as a type of "Web App", are similar to websites but built in a special way to enable more complex tasks. It's common to use these portals to collect and process information, automate tasks, or connect different systems together.

If you log in to different websites to do things, such as manage accounts or customer information, they are probably built as online portals. It's also possible to create an online portal that connects other portals together (using an API), taking the best bits from each and combining or automating tasks between both systems.

Mobile apps

If you have a smartphone or tablet, apps are simply the software that runs on the device. These devices have an app store where you can find and install software and tools. If your business is looking to make the most of apps, then you can create and publish your own apps to these stores.

Your phone contains all kinds of technologies, such as GPS trackers to know your location, accelerometers to know the device's orientation, internet access to read data from the internet, cameras to take pictures and scan QR codes, lasers to calculate the depth of objects, a screen to display the information, a microphone to listen and record audio, a Bluetooth chip to talk to other phones and devices... and the list goes on. Apps are the way to pull all of these technologies together to do something clever.

Servers, hosting, and remote databases

If you have to log in to a platform or app or store information or data that is accessible from lots of different places or devices, then these things need a server and a database.

A server is simply a computer connected to the internet that has a known address where people (or things) can connect to it. It's common to install a database on these servers, which can be used to store lots of information and all kinds of data, like a spreadsheet can.

"Hosting" is simply the process of storing your software on a server so it can be run and accessed over the internet. If you buy hosting, you are buying access to a server or part of a server shared with other people.

The Cloud

The Cloud is a fancy category of technology, software, and server solutions available on the internet. If something lives "in the cloud", it means there is a server (or group of servers) performing the core functionality of that thing. If you want to launch a tech idea that uses the internet, then part of it will inevitably be in "the Cloud".

It's common for tech companies to take servers, databases, and software and create their own specialised services that solve a particular problem. For example, Google has developed an image-recognition cloud service. It's just some code and databases on their server, but when you upload an image to it, it can tell you about the image — whether any people in the image are happy or not, etc.

Smart watch

This is a compact mobile device that attaches to your wrist. Like smartphones, smart watches can run apps, and these apps can make use of the various capabilities available from the smartwatch hardware.

Internet of Things (IoT)

IoT is the concept of connecting everyday items to the internet. If you take any device and put a mini computer in it, you can create software on that device that does something and sends information to a Cloud Server. This allows clever things to happen like being able to control the device remotely over the internet, or recording device data so you can see reports and metrics in an app or online portal.

Smart speakers

This is a minicomputer containing a microphone, speaker, and internet connection. These devices listen to voice input, convert them to instructions, and connect to services over the internet that perform the instructions, then use the speaker to tell you what has happened.

Machine Learning / Artificial Intelligence (ML & AI)

AI is a topic that makes most non-technical people switch off. It sounds so futuristic that surely there's no way you can take advantage of it, right? Wrong.

ML and AI are essentially just buzzwords for computers using statistics techniques to make a prediction. There are lots of techniques to generate these predictions and different ways those predictions can be used to solve problems.

If you plan to create something that will store a lot of data, it might be possible to use Machine Learning to make predictions using that data and display them to your customers.

The Google image recognition software, for example, read a lot of different images, got a human to say whether each face was happy or not, and used this data to create a statistics model of what means a face is happy or not (such as which pixels are dark relative to other pixels). Once the statistics model is built, you can put a new image into it without human input and the model will give a probability score between 0% and 100%. For example, if it returns a score of 60%, then this means the model is 60% sure that the picture contains a smile.

Maybe there are things happening in your business that could be automated with AI in a similar way, such as automatically categorising or labelling input data.

Technology bits and pieces

Imagine each of these technology types as different kinds of animals. You've got lions, tigers, and bears (oh my!), and each of these animals is uniquely suited to solving a particular challenge of nature. The tiger is made to hunt, with sharp teeth, while the giraffe has a long neck and teeth uniquely suited to eating plants, thus their features ensure their survival in the environment.

However, each of these animals are made of similar bits and pieces. Both the giraffe and tiger have eyes, a brain, a heart, skin, a cardiovascular system. Some features may be very different, for example a rhino has a horn and a zebra does not, but many of the principles behind what makes an animal can be similar.

The same is true for technology — although there are many different technology types, there are also many similar bits and pieces they can be made from. Here are some of the most common parts.

The front end

Often referred to as the client-side, the front end is the code that the user sees and interacts with. For web projects, the front end is usually the bits of the project that deliver the user interface, i.e. the elements that the user can interact with and see.

For smart and mobile device projects, the front end normally refers to the code that lives in the app installed on your device, which includes the user interface you see but can also include logic and functions that run on the device without the need to get information from a Cloud Server.

The back end

Information and data can be stored on the front end, but it must first be loaded from somewhere. The back end is that somewhere. Most modern tech ideas need a combination of front- and back-end technologies to work together. If your project requires someone to sign in or access information over the internet, then that information needs to live somewhere online. Otherwise, if you open a freshly installed app or web browser tab, how can it know who you are or what information you have?

The back end is the behind-the-scenes data and logic that sits on a server on the internet. Remember, a server is simply a computer that is accessible over the internet. So, in the same way we can write code for the front end, we can write code for the back end — code that sits on this internet enabled server.

You can think of the back end like the internal organs of a human, with a fully connected system that links your brain (the server) to all your organs to keep them functioning, while the front end is the skin, hair, and eye colour you can see when looking at a human. Then imagine that your brain isn't actually inside the body and is somewhere else, communicating telepathically to the rest of the body.

The framework

A framework is a standardised structure for front- or back-end code, and each framework is built using a particular programming language. Each framework comes with best-practice ways to do things in the code and automate or simplify common tasks and security considerations. As frameworks provide standards and techniques for the way people write their code, so they also make it easier to collaborate and ensure a high-quality product. In my opinion, the time saving and quality benefits of a framework mean that your development team should use one unless they have a very good reason not to.

The "app"

An app, short for application, is simply code that is packaged up so it can be installed and run on a device. This is what you download to your phone or view on your computer.

The "database"

Have you ever used a spreadsheet and had different tabs for different things with each tab containing many columns and rows of information?

For example, you can imagine a spreadsheet which contains a tab to store your business sales, with each column representing information about the sale, such as the price, item, and customer name, and each row representing a new sale.

A database is very similar to a spreadsheet and is designed to store lots of information of different kinds, but instead of tabs, you have tables of data, each with different column headings, and many rows or records.

When we design complex tech ideas, we can end up creating tens or even hundreds of different tables of data.

For example, a social media feed in an app might need a table of users, a table for the images they own, a table for the view or access history for each file, a table for which items each user has liked, a table for comments stored against each image, and so on. Part of designing your tech idea is your development team coming up with a plan of what tables are needed, what information they should store, and the relationships between the data in each table.

There are many different types of database, and your chosen developers will have to decide which type is best for the nature and scope of your project, considering what information needs to be stored, how much of it, and how frequently the data will be accessed.

The firmware

When building tech ideas, developers have different levels of programming languages they can use: high-level languages, and low-level languages.

If you were making a house out of toy building blocks, a high-level language would be like having a bunch of pre-made pieces you can attach together. For example, you may have a complete window or a set of roof tiles made of these blocks. This makes it faster to build the house as you don't have to make all of the different components of the house from scratch. Likewise with high-level programming, which makes it faster to build rich functions.

A low-level language is like building the house from raw building materials. It takes much longer but you have far higher levels of control over how each component is made, and you may be able to design things in a clever or more effective way, such as crafting a unique locking mechanism between pieces. There are many levels of low-level languages (such as binary and assembly), and some are so low level that it would be similar to being given raw plastic and having to melt the blocks together yourself.

If you have a small device, such as a smart camera, microphone, or sensor, this device can contain a small microchip or board that can be programmed. Due to the constraints of these devices, they can't run a full-featured operating system like iOS, Android, MacOS, or Windows, which means you can't install apps on them.

This limitation means that developers who want to program these chips must use low-level languages to create software that can run on these devices, which lets them do clever things like have full control over the sensors, inputs, and memory on the chips. This low-level microchip code is what we call the firmware.

Firmware development requires a different set of skills and way of thinking compared to the high-level code found in most apps. If your project requires firmware development, you'll find that developers who work with firmware all the time rarely build apps, and app and web developers rarely build firmware. This is similar to how you don't expect a bricklayer to also make the bricks. In the world of programming, high-level (apps) and low-level (firmware) development are different disciplines.

The "API"

An Application Interface, or API, is a way for two software systems to talk to each other. Think of it like a drive-thru fast food restaurant, where the two systems are the restaurant and the person in the car. The staff have been trained to accept certain commands such as, "I'd like a burger please" and you are conditioned to expect certain acceptable commands in response such as, "would you like fries with that?". But if you approached a drive-thru restaurant and asked if you could order a sports car with your burger, they wouldn't be able to help you!

The same is true with software APIs — there are a defined number of commands you can ask for and responses it can give. For example, your accounting software will probably have an API that allows you to ask "May I please have the line items for invoice 123?", and you might have another system that can automatically connect to this API, using the command to automatically read and do something with invoices.

If you have a tech idea that needs a server and a database, then you can also make your own API to make your customers, devices, tools or software able to read and insert information into your system automatically.

Creating an API between the front and back end is how we allow them to interact with each other, and this is what we call an internal API. It's internal because this API is only intended to be used by elements of the tech stack within your business. Internal APIs are a door with a lock that only you have the key to.

An external API is built in a similar way and uses the same approaches as an internal API but with your system explicitly designed to allow other people's systems to talk with them. External APIs are a door with a lock where you can give a key to other people to use.

Microservices

Microservices are a way to structure back-end software. So, rather than having one big, generalised software program that runs on one big server (known as a monolith), you instead make lots of very small, highly specialised programs that run on their own mini-servers. These mini-servers can be grouped and configured to automatically create copies of themselves if they are overloaded by the number of users or devices using them.

A microservices approach can be a great way to achieve millions of users or active connections to devices or things, but this comes at a cost as microservices involve more work to design and maintain, which can be cost-prohibitive if your idea is small or new.

For small- to medium-sized projects, or projects with cost constraints, a monolith or distributed monolith can do the job and is therefore the most common server type used on the internet today. For large projects that need to scale to millions of users or active devices, you may need to consider microservices where it is still split into separate services but each of those services is larger in scope. [44]

Visualising the technology stack

It can help to see how the elements of the technology stack interact with each other, so here is an example of a simple monolith structure, where there is only one back-end server running all of the back-end code:

Simple Technology Stack

Notice how all of the devices can communicate directly with the back end server over the internet but the Internet of Things (IoT) device can also communicate to the phone? This is because it's common for smart devices to have a little wireless module (often via Bluetooth) that enables them to communicate with the mobile app, which in turn relays information from the server to the smart device.

Usually, it's cheaper to have a connection from an IoT device to the internet via a phone as it requires less hardware on the device. However, this can have limitations, for example, the device is unable to connect to the internet without the mobile device.

If you hear developers talking about their tech stack, you'll hear them use terms like LAMP, MAMP, PyMP, MERN, and MEAN. This looks unusual and hard to grasp but it's just jargon for which server software they use. Each letter relates to a particular technology they've used on the front end or back end.

Each developer will have a preference for a particular set of technologies based on the size and nature of project, what they've worked on before, their training, or their education. You'll also find that every developer thinks their stack is the best stack and everyone who uses a different stack is doing something wrong!

Some technologies and tech stacks will have more tools available, some will have been around longer, some might have wider adoption in the development community, and some might be rising trends. The truth is that you can likely build your tech idea perfectly well in any of these tech stacks. The main thing, whichever stack is used, is ensuring that your developer has a good reason for wanting to use it, and that it aligns with the needs of your idea.

So make sure you ask the developer why they have proposed a particular stack, whether it's appropriate for this size of project, how popular the stack is in the development community, and whether there are any downsides to the stack. [45]

Closing words

Other notable mentions: augmented reality, virtual reality, smart glasses, haptic feedback, automated drones, smart screens, human brain interfaces, and body sensors. There are many different technologies and you can be sure that more will be invented, all of which you can leverage to solve problems and create unique value.

I'd like you to come up with several ways to solve each problem on your journey using one or more of the technologies mentioned in this chapter. Maybe your idea needs a back end and an app, or a front end and a cloud service, or a smart watch via an app.

Key takeaways:

1. You don't need excessively deep technical knowledge to implement a tech idea.

2. You can boost your creative thinking, effectiveness, idea creation, and problem-solving abilities by understanding the different types of technologies that exist and how they can solve problems in multiple ways.

3. Most internet-enabled tech ideas have a front end and a back end.

4. Remember to ask your chosen developers later in the process which tech stack they use and why.

5. There are different ways to host and set up your back-end server, each of which has cost and scale trade-offs.

The next two chapters are about the various ways you can make money from your idea and put a monetary value on any idea. If you're launching a product or service, both chapters will be useful. If you're building an internal tool, the chapter on putting a value on your idea will be the most relevant. But I want you to understand both principles because this will improve your ability to spot and decide which ideas are worthwhile. Together, these concepts will boost your business acumen more than they will individually.

CHAPTER 3:
WAYS TO MAKE MONEY
(REVENUE MODELS)

"Entrepreneurs don't usually fail from circumstance; they fail from what I call entrepreneurial rigidity – a fixed mindset and unwillingness to change the business model."
— Richie Norton

Chapter Relevance	
New Internal Tool	**New Product**
→	↑

"How do I make money from my idea?" is one of the most common questions entrepreneurs and businesspeople ask me at the start of their journey.

Let's face it, whether it's direct or indirectly, your idea will most likely aim to either make money or save money. Even if you're solving a problem for the greater good, it's possible to assign a value to the solution (which you will hear more about later). And even if you are a charity or not-for-profit organisation, these commercial lessons can be used to make sure your idea is self-sustaining. It's better to earn a profit from an idea that you invest or give away by choice rather than run it at an unsustainable loss that sees your idea have a short life.

This chapter covers the different ways you can directly make money from your idea. If you're not aiming to make money, feel free to skip to the next chapter... But not so fast! It may be the case that the tech innovation you're planning has a wider scope for improving your company performance than you initially thought. Thinking commercially about your innovation may also be a way to recoup some of the initial investment, contributing back into your departmental budgets.

There are many revenue model routes you can take, and identifying the right one for your start-up is mission-critical. Pick the wrong one and you could spend years struggling to make a profit; pick the right one and you should quickly start to see your revenue increase, enabling you to reinvest more into your idea and watch it grow.

The revenue model you choose can also heavily influence which features you plan and build, and when. If you don't consider the various ways to make money now, then you might find yourself wasting time and money reworking what you've built later.

But before we get started, here is a quick list of the 21 ways to make money from your tech idea:

- Method 1: In-app payments

- Method 2: In-app subscriptions

- Method 3: Software as a Service (SaaS)

- Method 4: Sell data or insights based on app use

- Method 5: Sell via partners

- Method 6: Free app with paid services (Freemium)

- Method 7: Affiliate commission

- Method 8: Ad revenue

- Method 9: Group licensing deals

- Method 10: Ecosystems

- Method 11: Product sales

- Method 12: Franchise models

- Method 13: Drop shipping

- Method 14: Enterprise solution selling

- Method 15: Pay as you go

- Method 16: Open-source premium support

- Method 17: Open-source ecosystem

- Method 18: Licence through partners

- Method 19: API licensing

- Method 20: Marketplaces

- Method 21: Auctions and reverse auctions

Method 1: In-app payments

In-app payments are one of the most lucrative revenue models for apps. They work by offering a (usually) free app to download, enticing users to spend money through one-off payments within the app. This seamlessly uses the credit card details the user has uploaded to their app store account. It also reduces friction to spend, making it more likely that users will pay for something in the moment if they think it has value.

A brilliant example of how in-app payments can generate revenue is Candy Crush. It was launched in 2012 and became an immensely popular game very quickly, with 250 million players still using it at least once a month at the time of writing.[46] When you play Candy Crush, you can go through it without spending a penny, but you can also pay for hints and help finishing difficult levels. This simple model led to the company making $86 million from in-app payments in December 2017 — their most profitable month.[47]

Gaming apps are best suited to in-app payments, but they're not the only industry that can generate revenue from this model. If your idea isn't for a game, you could offer additional features at an extra cost or keep certain areas of the app behind a paywall. For example, if you have an app that gives users access to articles about various professions or other unlockable in-app content or tools, you could add a small fee for users to read articles written by people who are well-known in their field.

Method 2: In-app subscriptions

In-app subscriptions are similar to in-app payments, but rather than being a one-off, users pay regularly for additional value. There are many well-known examples, including *The New York Times*, which has over 6 million subscribers, as well as Netflix and Hulu. [48]

The most popular in-app subscription is Tinder's premium access which is responsible for tens of millions in annual revenue. [49] Their model works by allowing anyone to use the free app without a subscription, with access to basic services. They then offer premium tiers called Tinder Plus, Gold, and Platinum. [50] These levels give users extra features designed to enhance the experience of the app, like being able to see who likes you and messaging someone before they've 'matched' with you.

The main benefit of a subscription model over a one-off payment approach is that you're guaranteed a regular income stream. By splitting a larger payment into smaller monthly chunks, it's also more appealing to users and extra features become more accessible. A user may not be able to spare £200 in one go but be happy to pay £16 a month.

However, this model does mean that your initial income is relatively low. Those first few small payments won't add up to much, and you'll have to wait to start seeing larger figures in your bank account, this can impact your time-to-money position, which is something we will cover in a lot more detail in Chapter 8: Raising and Managing Money. This is why you'll often see app companies offering a monthly subscription with a significant 20%+ discount to users who are happy to pay for a whole year upfront. Receiving a whole year of revenue upfront enables them to sacrifice some of their margin for the benefit of immediate cash flow.

Method 3: Software as a Service (SaaS)

Software as a Service (SaaS for short) is a cloud-based service model where users pay a fee to access software. Rather than downloading a program or receiving a physical copy of the software, users access it through an internet connection, removing the need to store anything themselves. This enables better accessibility and compatibility between software, and often requires a lower upfront cost than traditional models.

There are also plenty of perks from a developer's point of view. Because SaaS is based in the cloud, your development team can update it easily without affecting users' operations. The fact that it works across operating systems is another bonus. Rather than having to market separately to Windows, Mac, or Linux users, or having to build individual software for each system, you can build one program and market it to everyone.

Typically, SaaS runs on subscription payments, though you can also offer one-off payments at a higher cost that give users lifetime access. Many SaaS models provide easy scalability by allowing users to change their price plan to accommodate business or personal growth. For example, a business tool can be upgraded to offer more features and accommodate more staff logins.

You can also be creative in the way you bill for SaaS platforms, for example, you might charge a fee based on the number of users, downloads, or data, or the level of access to the platform and its features. If you can identify where the value comes from in the feature set, you can align your price to this value. This approach is why per-user pricing is so popular — if a business needs more users, they are probably getting enough value per user to justify the extra cost.

The world of SaaS is vast, and if you can create an innovative solution to a problem using this model, then you can certainly create revenue from it. [51]

Method 4: Sell data or insights based on app use

It's not just your users who can add to your revenue — there are also outsiders who will find your app indirectly valuable. One of the most popular ways to generate money from non-users is through the sale of data or insights. When someone uses your app, they can consent to make certain data accessible, including their location, contacts, and other information. You can then send this data to third parties. Though you must be careful with personal data as it might put you in breach of data protection laws; if in doubt, get permission and anonymise the data.

For example, location information is often useful for marketing agencies. If they can tell where you are, they can create more targeted and appealing ad campaigns for you. If you live near a fast-food chain, you might be more enticed by their deal than someone who doesn't live as close. Or if you work from home most of the time, you may see more ads for home office equipment. For a marketing agency, this sort of information is invaluable, so they'll pay good money for it.

In fact, this revenue model is so popular that one study, which analysed almost a million apps on Google Play Store, found that the majority had enabled third-party tracking and was selling data to a vast list of huge global corporations, including Amazon, Microsoft, and Twitter. [52]

However, you do have to be very careful if you take this route. There are a lot of policies surrounding selling user data to third parties. Both Google and Apple prohibit selling data unless it's related to improving your app or the ads displayed within it, and they are constantly tightening their data-tracking rules. [53] [54]

You should also be very careful about the morality of collecting and selling data. To decide if your use of data is ethical or not, ask yourself this: if the mainstream media published how you were using customer data, would you be happy that you could defend yourself without causing reputation damage?

As such, you must always gain consent if you track users. Note that if users have to allow data access in order to use your app, this can reduce your market share. But if you give them the choice, then be prepared for many users to not give their consent. In the US, a recent study at the time of writing showed that 96% of iOS users had opted out of data tracking. [55]

Method 5: Sell via partners

Selling via partners is logistically quite different compared to selling directly.

Selling directly refers to buying from the company or individual who developed the product. For example, if you buy an Audi car from an Audi dealership, then Audi sells that car directly to you. When you don't buy from the developer themselves but from a third-party supplier, this is known as selling via partners, like buying an Audi from a generic car dealership.

Accounting software Xero is an example of a company that does both. [56] They sell directly to customers but also sell via accounting firms, who are therefore their partners. The accounting firms build relationships with Xero and get the software at wholesale price, offering it to customers of their accounting services and getting a small commission.

In this example, it seems like Xero is selling at cheaper prices so that firms can make a profit... What's the point of that, you might wonder?

Marketing nearly always carries some cost and paying a commission to a partner can be seen as just another marketing expense. So, some providers will opt to spend nothing on advertising and allocate all of the marketing budget to paying a commission to incentivise key partners instead. [57] What businesses care about is how much it costs to acquire a customer relative to the amount of money the customer makes in revenue and profit.

This partner relationship therefore makes sense because the customers of the accounting firm are the exact type of person that the accounting software aims to reach with its services, and customers who use modern accounting software are generally easier for an accounting firm to work with than those who don't. The relationship is therefore in harmony, and there are many benefits to all of the stakeholders involved (the customer, accountant, and software provider).

Working with partners can help you to increase revenue and reduce marketing costs. However, if you sell via partners *and* directly to customers, be careful not to undercut your partners or they may stop selling your product.

Method 6: Free app with paid services (Freemium)

Freemium revenue models are where you provide free access to a basic level to your product or service so you can deliver value, engage the customer in your way of doing things, and use their loyalty to up-sell them your premium offering.

The Mailchimp email marketing platform became a runaway success by implementing a freemium business model.[58] It allowed users to sign up and use its platform for free, sending email campaigns to a list of subscribers. The hook? Users were only allowed up to 1,000 email subscribers on the free version, after which they had to pay for premium.

The logic of this model makes sense for Mailchimp. If someone is using it extensively and builds a list of 1,000 subscribers, then they are clearly getting lots of value from the service. Reaching 1,000 subscribers means they have likely invested significant time and energy in nurturing their list and creating campaigns on the platform. So, when they reach the subscriber limit, they don't want the time and hassle of moving to a different platform — it's much easier to just upgrade.

The freemium model makes Mailchimp's services sticky! This means that although Mailchimp has a lot of free users, they are able to charge more per email or subscriber than platforms that don't operate this model. Why? Because those who go on to pay are often happy to pay a bit more for the benefit of convenience.

By offering a free app or SaaS portal with paid services, your users aren't forced to commit to you. However, you get the chance to show them what you have to offer, gain their trust, and convert free users into paying customers.

Method 7: Affiliate commission

Affiliate commission is the method of advertising another company's products or services on your app (or platform) in return for a small commission when someone clicks on the link or buys something from the company. For example, Microsoft have an affiliate program for their Office tools in business accounts. [59] If you show their ads or promotional content in your app, you can earn revenue every time someone clicks on the link and makes a purchase. The highest commission on a single purchase is $20 but your commission increases for multiple users (for example, two accounts = $40, three accounts = $60, and so on).

Affiliate commission models are similar to partner selling but the relationship between the solution provider and partner isn't as close and collaborative; they tend to be more transactional. The growth of the internet and e-commerce has allowed affiliate models to grow rapidly. [60] It's now a well-known way for people with a platform to earn some extra income.

Microsoft are just one of many companies offering affiliate programs, which gives you the opportunity to earn a fair amount if you work with multiple companies. From ad banners to simple links in a blog, you can build up a diverse affiliate portfolio and start creating passive income streams. To maximise your earning potential, find companies and products that your audience are likely to be interested in — but don't saturate your app with affiliate links or you'll risk losing your audience's trust.

Method 8: Ad revenue

Ad revenue models are where you display adverts on your site, platform, or app and make money from the companies that pay to display those ads.

Many people I speak to believe that generating ad-revenue is the holy-grail of making money from their tech idea. Their thinking is "if I build it, I can give it away for free and see the money roll in when people start to use it." In some ways, they are right — after all, some of the biggest tech businesses in the world make the majority of their revenue from selling ads in one way or another. [61] However, the reality is that creating a business which makes all of its money from ad revenue is super-tough.

Let me put this into perspective. Content creators on YouTube aim to build a sizable audience and generate millions of video views so they can make money from their audience. While many Youtubers offer side products and services to earn revenue, the main way many of them make money is from people clicking on ads in the videos.

At the time of writing, the average YouTuber can expect to make between $1,000 and $3,000 per million video views — depending on the audience they target. For example, finance videos are higher-earning while kids videos are lower. Either way, think about that figure. *One million views is only worth $3,000.* If your platform or app is new, consider how much work it will take to generate a million views just to make a few thousand.

Businesses that make money from ads have generally spent years of time and money growing their platforms and building their infrastructure. They've reached a level where enough viewers come back each day for their view-count figures to mean that they can sell ads profitably. That's years and years of patient hard work with zero profit to reinvest. If you haven't raised mega-bucks to get your idea off of the ground yet, then the ad-revenue business model is going to be very challenging.

That said, I don't want to put you off of the idea of ads completely as there are ways to build your audience organically to achieve ad sales. For example, you might already have a strong audience or be an influencer in a niche market. In this case, launching an ad-revenue model may be an easy way to make money from your established audience.

Method 9: Group licensing deals

Group licensing is when a group of users or companies buy usage rights to your software or app for a set number of users, usually at a discount. It's commonly used by businesses who need multiple employees to access the same app, families with the same needs, or a wider membership group where the businesses want to pool their members to get good discounts.

For example, the Federation of Small Businesses in the UK represents hundreds of thousands of small businesses. If they approached you asking for a bulk-deal for all of their members to access your tech idea, you'd probably be very interested. Why? Because although the per-user revenue might drop significantly, you can close a deal worth 100,000 times that of selling to customers one by one.

There are group licensing deals aimed at consumers across industries. The music steaming service Spotify, for example, offers a range of premium group plans, including families and businesses. [62] This allows one family member (a type of user group), to sign up and provide access for the rest of their family.

If you can determine a logical group of users or companies where a small group of people (or one person) represents the interests of that group, there may be an opportunity to create a special pricing model so this group chooses you over the competition.

Method 10: Ecosystems

Ecosystems revenue models are where you aim to make money from the combined value of the products and services that you offer, rather than relying on any one method. Ecosystems mean that that your tech idea doesn't have to be a direct money-maker, and sometimes it can make commercial sense to develop a new product or service simply due to the value it adds to everything else you do.

Revenue streams can branch off your product in an ecosystem that combines complimentary software, devices, products, and even other applications. This is where one or more of the services you provide (such as an app, portal, or IoT device) becomes a jump-off point for users to delve into the rest of your ecosystem and spend money with your company.

The key thing in an ecosystem is to make sure everything connects with everything else, i.e. that the value of one product or service aligns with the others.

Apple, for example, has become successful by growing its ecosystem. Customers who purchase an iPhone buy into the brand and status of the company. They become an Apple enthusiast and also want the Apple

Watch, speaker, MacBook Pro, and AirPods. Apple knows that by having an ecosystem, they grow the average lifetime value of each customer, as over time (probably years), these customers will loyally buy many or all of their products and services.

There are many ways to create an ecosystem. For example, whether paid or free, you might start with the creation of an initial app, platform, or tool. Over time, you can grow the number of users who receive value from this tool, then expand the capabilities or services in your ecosystem to make money from this audience. The app acting as the hub in the middle pulls the strategy together. When you think of your tech idea as a marketing tool within your ecosystem, the entire concept of making revenue from your tech idea shifts.

Ecosystems are powerful, and while it may not be appropriate to create an ecosystem at the beginning of your journey, you should certainly be thinking of ways to gradually build one, even if it takes years to achieve. We talked earlier about aligning decisions around your vision — and maybe part of that vision can include considering what your product and service ecosystem might look like in three to five years' time.

Method 11: Product sales

Product and service sales are where you make something and sell it directly, taking payment at the point of transaction. If your tech idea aims to enable this type of direct sale, then this method is for you.

If you have the right products, your tech idea can be your shop window, such as a website or mobile app. In fact, there are many companies nowadays that sell their products entirely online, with no physical brick and mortar shop at all.

Online clothing giant ASOS is a prime example of this modern tech-sales-only product business. [63] Just 20 years ago, it would have seemed impossible to completely operate a clothes retailer online, without the option of visiting a physical shop. Common reservations would be: *How would customers try on clothes? Who would buy something if they don't know it will fit them?* Yet ASOS has reached billions of revenue, and all without a single shop that you can visit in person. [64]

The opportunities to sell products through apps are undoubtably there. Whether you're selling something physical, like clothes on ASOS, or something digital, like a piece of software, you can do everything through an app. Payments are more secure than ever and there are plenty of ways you can keep customer data safe, almost entirely removing risk for the customer when done properly. The capabilities of apps makes showing off products easy, and the number of people shopping online is on the rise.

Indeed, over 75% of people buy something online at least once a month. [65] If you have an app *and* something to sell, why not combine the two and start reaching that huge audience?

Method 12: Franchise models

Franchise business models are where you create a proven way of operating a business, from sales, marketing, branding, processes, and procurement all the way to fulfilment and customer services, then you sell the rights to independent investors who buy into and set up their own instance of the business model for themselves.

McDonalds, [66] Costa, [67] and Toni & Guy [68] are all franchises in the UK. They've built up a set of business processes to create a wider model that works well, and now they can sell that model on for a profit.

When you're creating a business, you have a lot to learn. Let's say you were opening a coffee shop — you'd have to consider table placement, where to source your coffee from, what till systems to use, amongst a tirade of other things. A franchise lets you skip this process by paying to use a model created by an existing brand, taking the decision-making out of the process. It also often gives you a name that people already know, which is a huge head start in marketing. If you were to open a Costa in any town in England, it's likely that most of the inhabitants would know your business.

Franchisers recognise the value within their own processes. Outsiders are willing to buy into the franchise because the business has already proved itself — and the brand recognition they've garnered will lead to more sales for the individual. When the franchisee buys into the company, the franchiser gets a lump sum to use the business model and can even ask for a cut of the profits as the business moves forward.

This franchise model can easily be transferred to the world of software and tech. If you have a tech idea with a lot of value and a process that's working, becoming a franchiser is a brilliant way to scale your business. However, you do need to prove your concept from the get-go and ensure it can work as a franchise model, otherwise you might struggle to convince investors that it's worth buying into and setting up a new business.

Automating your processes is simple in software and is a step towards showing your value as a franchise model. Having a software platform that makes your franchise model possible and delivers unique value can be core of the value proposition in this business model.

Remember that software is one of the most powerful ways to automate and systemise processes. If franchisers value and want to buy businesses whose processes are robust, then software can become a huge part of why they might choose to invest in your franchise.

For example, you could create an app that automates business processes, such as sales analysis, cataloguing products, and integrating with third-party systems. Franchisees have a login for this tool and can easily run the business model from here. This digital asset generates buy-in interest and becomes difficult to disconnect from, solidifying your place in the franchise industry. Once vendors use this platform, it becomes sticky and difficult to move away from, delivering competitive advantage to your franchise.

Method 13: Drop shipping

Drop shipping is the process of selling items for a third party through your own platform or app. Rather than buying products to sell, storing them in a warehouse (or at home), and sending them out yourself (fulfilment), you simply make the sale and let the third-party provider do the rest of the work.

Here's a simple overview of the drop shipping process:

1. You find products you wish to sell from a drop shipping supplier.

2. After getting their permission, you advertise the products for sale on your app, often creating an automatic integration between your platform and theirs.

3. A customer orders one of the products from your website, portal, or app.

4. Your system automatically sends the order details to your supplier, who sends it to the customer.

5. You pay the drop shipping supplier their cut if you took payment, or they pay you if they took payment.

As you can see, you never have to come into direct contact with the product to sell it. You make a profit based on the difference between what you paid the supplier for the product and what the customer paid you. For example, if you bought a range of mugs for £3 each and sold them on at £6, you'd have a £3 profit for each.

So, why would suppliers agree to this? Here comes the tricky part.

To start seeing sales, you have to put in a lot of work — or money if you're hiring a tech team. And by this point, you should be aware that making an application is no easy feat! Once you have the app, you have to market your business, make sure people can find your products, and compel them to buy those products. The supplier doesn't want to do that — they simply want to sell the products at the flat, wholesale price they're asking for and are happy for you to sell them with your markup attached if you're willing to put in the work.

Thanks to the likes of Amazon [69], Shopify [70], and other dominant online marketplaces, drop shipping's place in the world of revenue streams is well and truly solidified. As a method of passive income, drop shipping has become synonymous with entrepreneurship and a digital nomad lifestyle, letting you earn money wherever you are in the world. Of course, it's not always as easy as it sounds and a lot of effort goes into setting up a successful drop shipping company but if you get it right, then you could be in the market for a lot of money.

For example, NotebookTherapy is a niche Japanese and Korean stationery shop that follows the drop shipping model and has been very successful, with over 1.4 million Instagram followers and thousands of likes on every post. [71] They have over 800 reviews on Trustpilot (with an average rating of 4.4) and they rank top on the first page of Google for the search term 'Japanese Stationery' in both the UK and the US. [72] They're one of many companies that demonstrate how drop shipping can be an incredible revenue model, but there's no doubt they've worked hard to get there.

If you have marketing and sales skills but don't want to go create the tech product or service yourself, then drop shipping can be a low barrier to get started and make money quickly. However, it is hard work and competitive, and you might need to operate on wafer-thin margins, so it's not a business model for the faint-hearted.

Method 14: Enterprise solution selling

When I say enterprise, I mean big business. Enterprise solution selling business models involve you creating a product or service that is highly valuable to a small number of very large customers, where the sale price of the product is high but so is the relative value it brings to the target customer. An enterprise solution model looks to solve a complex problem for an entire organisation rather than for an individual.

An enterprise solution is usually tailored to the business you're selling to, catering to their exact needs rather than offering a generic product. It's then offered as a license, allowing you to control how they pay for and use the product, usually with upfront costs to install or develop it, followed by a recurring management or service fee.

Because an enterprise solution model looks at personalised tools for businesses, your company will have to engage with your potential customers very deeply to understand their pains and propose a tailored solution. You'll need to identify their needs, build a relationship with them, and create a product using the information you've gathered.

The downside to enterprise solution models is that it can take a long time to identify, build relationships with, and close sales into the target businesses. However, the benefit is that you're hunting whales, so a successful sale could be worth tens of thousands, hundreds of thousands, or even millions of pounds in revenue.

Also, depending on your tech idea, this model enables you to sell the *concept* of the project to the client, where it's expected that there will be a build phase before the technology is ready. I've seen clients sell a product or service they've not yet built and cover some or all of their development costs from just that first client. The speed to which your business becomes revenue generating of this business model can be excellent, though your sales skills and credibility also need to be exceptional to have any chance of closing this scale of deal.

Method 15: Pay as you go

Pay as you go is a revenue model where you only charge customers for access to your services at the point of need, and the charges are proportional to the amount of use.

Early pay as you go models became popular with the mobile phone market exploding in 1997, and they really started to make an impact in the early 2000s.[73] The idea was to buy a phone and a separate "pay as you go" SIM card. You could add money to the SIM card as and when needed and weren't tied into a contract. This approach made mobile phones far more accessible to a lot of people.

The key with your tech idea is to understand your idea's value proposition and whether the pay as you go model enhances that, which often depends on your demographic. For example, pay as you go SIM cards were such a success because they appealed to younger phone users with less money, such as teenagers, or those who didn't use their phone often. They didn't have to worry about a contract or monthly bill, giving them more freedom — a significant unique selling point (USP) to companies who chose to provide these services.

Method 16: Open-source premium support

This method involves providing access to your software source code for free and making money from supporting services aimed at the business users of that code.

Open-source software is often free to use, and users can download and make edits to the original source code. Developers who build open-source software are usually part of a community and are happy to build elements for free — knowing that others do the same. All of the contributors benefit from something they have worked on together, and everyone writing code for free can benefit from the free code written by others.

Although the code is free to use, there may be complexities to the process of using it. For example, what if the server running the mission-critical code goes offline at midnight on a Saturday? Businesses want to mitigate these risks by appointing trusted service providers to manage and maintain these services, even if the code itself was initially free.

With these kinds of support companies, their customers are commercial businesses that benefit from using the software but also want to do so in a secure and robust way. They want issues to be resolved quickly and their systems to be up to date with the latest software versions. These companies can't afford to do without configuration, security patches, and maintenance — so they'll pay for it.

Method 17: Open-source ecosystem

This method can be used in combination with the open source and ecosystem methods mentioned already but is subtly different as the main goal with this approach is to achieve mass adoption of your platform. Once you achieve this, you can use it to reach users in new and creative ways. A lot of companies create whole ecosystems around their central code, which creates a loyal customer base and increases profits.

Google's Android operating system (OS) is a prime example of this. Google bought Android for $50 million in 2005, expanding their empire and branching into an area they didn't previously have expertise in. [74] The Android OS is an open-source operating system for smart devices built on Linux, so Google bought something that is widely and freely available. However, they saw value in the ecosystem, knowing they could build around it to support their main revenue models, driving ad revenue from searches.

Today, there are billions of Android devices in active use all around the world. [75] Each of these devices is cram-packed full of Google software, including their search engine, which is the most lucrative source of revenue for Google. [76] They also come with the Google Play app store, which brings in billions in revenue. [77] It's easy to see that the value of buying Android wasn't in the software itself but the ecosystem they could create around it.

If you create a piece of software, consider whether you can find more value by open sourcing it. Open source may bring you a larger initial audience who will buy into your ecosystem the more they use your code.

Method 18: Licence through partners

This revenue model involves another business adding value to their product or service (which their customers buy from them) but where part of the value they sell comes from you. This is similar to how your car is made from many parts that were once manufactured by companies other than the brand of car.

With this revenue model the user is not always the buyer, and the buyer is not always the user. Like children's toys, it's the child who wants and will use the toy, but it's the parent who makes the choice to buy it. When you're selling a product or service this way, you need to consider how the value proposition of your offer relates to all stakeholders in the chain.

It's common in the software world that the end customer using your product is not the buyer, and the buyer is the software developer they've hired to create a solution for them.

Cloud technology services provider Amazon Web Services (AWS) follows this licence to solution providers revenue model. [78] AWS offers a range of services aimed at developers who are implementing solutions that make technical work easier and enable complex functionality.

AWS includes services such as creating scalable databases, machine learning tools such as facial recognition, or setting up Internet of Things (IoT) devices to communicate with third parties. If every developer built these capabilities from scratch, it could take hundreds or thousands of person hours. Amazon realised this and implemented a solution that allows developers with similar problems to make the most of pre-made services and deliver rich functions far more quickly for their clients.

In this relationship, although the developer is using the services, it's the developer's end customer who pays. As such, both need to be involved in the sales process and both need to be convinced of the value that the services bring relative to their cost.

If you make a service or code that adds value to solution providers in this way, then you can control how you price it and what the customer can do for that price (i.e. the licence terms). The license could be a recurrent subscription (monthly, yearly, etc.) or a one-off payment for the current version with extra payments for upgrades. [79]

The more developers who work on the project, the more licences you'll need. Some businesses also offer a free non-commercial license to encourage non-developers to use their software and buy into their ecosystem. So, this revenue model can flexibly incorporate a combination element of the other revenue models we've discussed so far.

Method 19: API licensing

API licencing is where you create and sell access to a digital service, which other developers can programmatically connect to and use as part of their software solutions.

Earlier, I spoke about APIs — application interfaces that allow two systems to communicate with each other to transmit data and automate tasks. External APIs allow clients to directly interact with your system, and API licencing is all about how you make money from creating and selling access to your external API.

If you can solve a complex tech problem and make it accessible through an API, you can license it to others. The basic idea is: the user inputs information, your software solves it, and outputs a conclusion based on that information, which the user benefits from.

For example, language APIs exist to translate between languages. Developers integrate with these APIs to automatically translate text and pay per use or per thousand uses. Though there is a cost per use, this can work out much cheaper than building a translation tool from scratch.

Once you have built your API, you can licence it for a fee to make money. The industry standard approach for this business model is to bill each time the API is used to do something (called "API calls"), though you can also bill for use in other ways.

Method 20: Marketplaces

A marketplace business model is when you have a buyer and a seller, and your platform or app facilitates the sale, taking a cut in the middle.

A lot of the tech ideas I've come across in recent years involve some form of marketplace business model, probably because people are familiar with using marketplaces, appreciate their convenience, and find it easy to spot opportunities where a marketplace could improve the way customers buy in a particular sector.

Uber is a well-known example of a marketplace business that connect users to taxi drivers and provides the whole infrastructure around that process. [80] There is also Amazon Kindle for digital products, which connects authors of digital books to customers who like to use e-readers. You can probably think of many more.

Though marketplaces can add a lot of value, it can take a long time for these ideas to make enough revenue to cover their costs, let alone be profitable. This means that for marketplace business models, you need to be patient, be able to finance a suitably long runway to build, test, and release your technology, and not expect immediate returns. [81]

Method 21: Auctions and reverse auctions

Auction applications are a form of a marketplace business model. The concept of the auction has been around for thousands of years and has been popular for decades.

Of course, the most famous auction site is eBay,[82] which was founded in 1995 and has since seen its share price grow 100-fold.[83] Since their launch, plenty of other auction sites have become incredibly popular to capitalise on this lucrative business model.[84]

The vast majority of auction apps work by allowing users to upload their own products, and other users can bid on them. When a sale is made, you take a small fee as the third party who connected the seller to the buyer. It's a simple revenue model and doesn't require you to sell anything to your own customers.

There are also reverse auctions. For a developer, the two are very similar. The difference is that the seller uploads the service or product first, and those looking to sell the same thing place bids. The lowest bidder gets to sell their goods.

Both models are incredibly easy to set up and, with a well-designed app and quality marketing, they can become stable revenue models.

Closing words

Those are the 21 ways![85] Make sure you carefully pick the ones that work for you as this choice will heavily influence the decisions you make throughout your journey.

I imagine that a handful of options jumped out at you and felt instinctively right. If so, I recommend writing them down and listing the pros and cons of each method relative to your objectives.

For example, if you think the best revenue generation approach is to display digital adverts, then this may allow you to acquire users more quickly, but it may take you a lot longer to make money from those users.

For the moment, keep the various options in your back pocket. Save the final decision until later. In Part 2: Deciding Factors, I'll give you the baseline knowledge and tools to pick a winning approach from this seemingly endless range of options.

Key takeaways:

1. There are lots of ways to make money from your idea, each with different implications for every aspect of your business.

2. The way you market your idea can have a big impact on which method of making money is best for you (more on that in our bonus Marketing chapters later).

3. Regardless of which approach you take on day one, you should consider how to create an ecosystem as your business grows.

4. Marketplace businesses, though commonplace, require a lot investment and time to establish, which might have strategic or fundraising implications.

5. You should determine which revenue model is best for you by considering who the real buyer is: the end user or another interested party (such as a business that wants the data or to market to the end user)?

CHAPTER 4:
IS YOUR TECH IDEA WORTH EXPLORING FURTHER?

"It's easy to dream, but much harder to execute it."
— *Gary Vaynerchuk, Veyner Media*

Chapter Relevance	
New Internal Tool	**New Product**
↑	↑

Before we run ahead and begin planning which features should go into our tech idea, we must first decide whether the idea itself is worth exploring further. I know that by this point, you might be excited to run ahead but it's important not to waste time on ideas that have no prospects. The problem you address must exist, and be large enough to be worthwhile.

We're going to score the value of your tech idea using a special point system. More points mean the problem is more worthwhile to solve. Fewer points mean the time and resources to solve the problem aren't worth it...

This point system is money!

In this chapter, we'll cover ways to put a monetary value on your idea, and we have some tricks up our sleeve to do this for ideas that don't make money directly through sales and revenue. This is because every problem that exists has a hidden number behind it, which is the cost that this problem introduces to the world, and we can make some clever guesses to estimate what that number might be.

Putting a value on the problem you aim to solve helps to put things into perspective.

For example, is it worth implementing a complicated solution to a problem that costs the world £10 per year? What about £10,000, or £10 million? The larger the cost of the problem you aim to solve, the more worthwhile it becomes to address that problem.

Putting a value on your problem ensures you spend your precious time doing the right things, deciding which ideas should be dropped and which should be explored a little further.

Some methods of judging value are obvious — if your idea saves money, you can simply compare the cost to develop the idea against how much money it saves. There are also some less tangible measures of value, such as if your idea creates something that improves the prominence of your brand.

<u>A quick warning if you don't like maths...</u>

Unfortunately, we can't make successful business decisions if we don't talk about numbers and money at some point, and this chapter is that point. I'm about to show you different ways you can assign a monetary value to your idea, which means we need to add, subtract, multiply, and divide some numbers. So, you might want to skip over the examples, get straight to the lessons, and come back to the examples when you have the time and brain power.

Each of the number examples are a box like this:

> **EXAMPLE**
>
> $1 + 2 = 3$

We can work out the value your idea brings with seven back-of-envelope tests. These tests are quick methods to figure out some numbers and multiply them to get a rough value estimate. The sums you perform should be super simple, so simple that you could scribble them down in a few seconds on the back on an envelope, hence the name.

Test 1: Does your idea save time?

Time saved can be yours or your client's. Time that you can save yourself should be considered a cost saving, and time you save for your client should be considered a form of value-generation.

But how are you supposed to work out the time saving that your idea brings?

It can help to think about the processes your business has or will have, then walk through the steps in those processes one by one.

A sequence diagram is an easy way to identify the different parts and people required to make sure a given process works. These diagrams are a visual representation of the steps. Outlining processes in this way can help you understand the process better at a high level, who is involved when, and where the process can be more effective, such as reducing the time at each step or spotting new opportunities to reach customers.

To draw a sequence diagram, start by drawing a line for each stakeholder or relevant part of the process. Let's imagine a simple process that involves your business and your client's business. We could split these further and add lines for each department in each business. You should use your judgement to decide how many sequence lines make sense to model your process: [86]

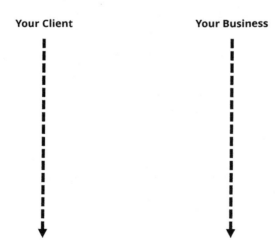

Each of these arrows represents parallel timelines for both you and your client. Dots parallel to each other happen at the same time. For example, this is how you could record three separate unconnected events:

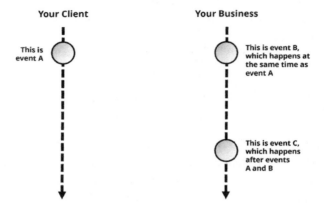

You can take this one step further by drawing lines between the dots to indicate which events trigger other events.

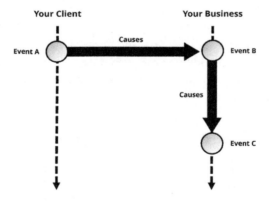

In this example, event A triggers event B within our business, which leads to event C.

Let's now apply this technique to a real-world example: the process of receiving a telephone customer enquiry from your website. This is how you might sketch the user journey of a client visiting your website, finding your phone number, and calling your customer services department:

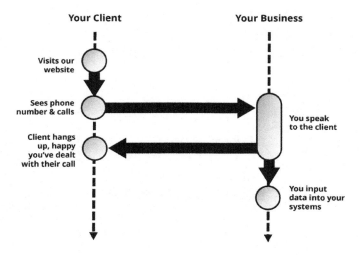

Notice how one dot has been elongated to show that it has taken some time to happen?

If you have a more complex process, then you may need to draw additional lines for the various stakeholders in your business, third-party tools, or anything that takes your fancy. Don't worry too much about conventions — what's important is that you understand it.

Once you've mapped out of each of the steps in the user journey, you can assign average times to each dot:

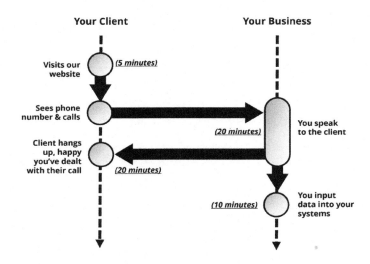

This diagram shows that our current process takes up 25 minutes of the client's time and 30 minutes of our own. Now, let's see how this process could look if we replaced the telephone number with an online offering:

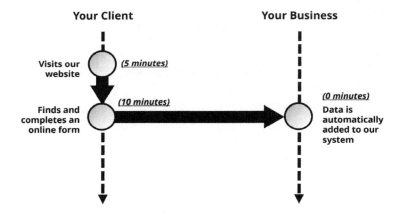

Not only has our new idea reduced the client time by 10 minutes, but it has also eliminated 30 minutes of admin time in the customer service rep being on the phone then manually inputting information into the system.

Remember, this sequence diagram is supposed to be a back-of-envelope test so you shouldn't have to spend more than a few minutes drawing out your diagrams. For simple journeys, you may be able to do all of this in your head.

Now you can scale these numbers to work out the savings across an entire month or year.

Example

If the average client call is 20 minutes and the new enquiry process drops this to 10 minutes and you receive 100 calls per month, then you will save 1,000 minutes of client time per month:

10 minutes x 100 calls = 1,000 minutes per month

From your business' perspective, the original process took 30 minutes to speak to the client and input the data into your system. The new process drops the person time involved to 0.

This is a full 30-minute saving across 100 calls per month, representing 3,000 minutes admin time saved for your business.

30 minutes x 100 calls = 3,000 minutes per month.

That's 200 hours of client time and 600 hours of administrative time each year.

3,000 minutes x 12 / 60 = 600 admin hours

However, be cautious trying to strip every second of time from every process you find as faster doesn't always mean better.

Some businesses may opt for a slower more personalised service as part of how they differentiate themselves and add value for their customers. If you ignore this business objective, you might risk stripping time from a process to save its value but costing more value elsewhere in the business!

If you can trim person-time from a process whilst eliminating any downsides, then you make your business leaner, more competitive, and more scalable.

Armed with this sequence diagram technique, you should be able to sketch even the most complex processes in a matter of minutes. Use it to spot opportunities for streamlining your process using technology. This could be for a process you already perform in your business or a process you see other companies following.

Test 2: Does your idea save lost opportunity cost?

In the previous test, we measured the time savings, but wasted time has both direct and indirect opportunity costs too.

A direct cost is what each unit of wasted time costs to you to buy. The indirect opportunity cost is what that hour is worth to your business.

Example

If your employees cost you £40 an hour, and you lose 600 hours per year, then the direct cost is easy to calculate as £24,000.

$$£40^{\text{hourly employee cost}} \times 600^{\text{wasted hours}} = £24,000^{\text{direct cost of wasted hours}}$$

Though the direct cost per hour is easy to calculate, the opportunity cost is often more important as it represents much higher levels of waste. And every area of waste you can fix is an opportunity to be more productive.

Every hour saved completing a task more quickly is an hour that can be allocated to a more useful or productive task. The direct cost per hour of an employee might only be £40, but if you bill clients £100 an hour, the value of each hour saved is £100, not £40.

Here are some more examples of different kinds of opportunity cost savings:

The opportunity to complete more billable work

Imagine you couldn't win a new £100k project because you didn't have enough staff available in your team to complete the project — they were doing other things instead. This £100k would be the opportunity cost of not implementing your time-saving idea. Think of the additional revenue you could make if you could stop those staff wasting time on unproductive activities.

What if you could more productively allocate your team so that you are able to win this £100k project without having to recruit more people?

Or what if you have a business where the customer service's telephone number is answered by the sales team rather than by a customer service representative?

Every minute that the salesperson is dealing with a support request is a minute they aren't spending productively drumming up sales. A month of lost salesperson time could be extremely costly to a business when measured in terms of the opportunity cost.

Example

If each sales representative has a £50,000 sale target each month, then five months of lost salesperson time is worth £250,000 to the business each year:

$$£50,000 \times 5 \text{ months} = £250,000 \text{ lost per year}$$

When multiplied across an entire team, this could be a very costly activity. If you had 10 people in your team, then the opportunity cost lost could be well over £1 million:

$$£250,000 \times 10 \text{ people} = £2,500,000 \text{ lost per year}$$

There is a lot of money to be made in this example by ensuring that your highest value-generating employees are working on the things that are worth the most to your business. And if you're a small business founder, this opportunity cost of your wasted time is greater than the wasted time of your employees.

If you're a business founder, your weekly activities will be split across a wide range of tasks, such as sales, marketing, HR, accounting, admin, operational activities, and planning. You may have less than half of your week available to dedicate to these high-value sales activities. This means that saving time on admin to free you up to sell more could have a huge impact on the success of your business.

The opportunity to reduce service costs

There are many ways that digital systems can eliminate the need for you to use third-party services, suppliers, or products. For example, if you can replace the process of manually posting letters with digital ones, then you could save money on postage costs.

It's also common to pay for software per user per year. If you have a big team, say 100 people using a system that costs £15 per user per month, that's £18k each year or £54k over three years. In which case, it may be cheaper to build your own internal software.

If your tech idea delivers economic or environmental improvements in some areas, then you may be eligible for grants or tax rebates. If your tech idea is pivotal in enabling you to benefit from such grants or rebates, then their value counts as the value that your tech idea brings.

The opportunity to avoid legal battles

If automating a process reduces the likelihood of human error, then there may be certain legislative benefits. For example, if you make errors less frequently, then you may avoid costly legal cases where people sue you due to those errors.

On average, large businesses spend £1 million on litigation each year, and even for small businesses, the average civil case costs of £50,000–£100,000. [87] That could be a business killer, so it's worth thinking about.

The opportunity to improve employee focus

Sometimes, your tech idea may make it easier for people to focus their time on high-value opportunities over low-value ones. Often, efficiency is achieved through data and insights that help you decide what to prioritise or do next.

For example, imagine a salesperson who has more people contacting them about products and services than they have time to handle. How should they decide who to contact first? If they focus their time on contacting low-value prospects over the high-value ones, they could be losing a lot of money in sales. But if they had a tech tool that scores prospective customers based on how likely they are to buy, then with the same day-to-day effort, the salesperson could prioritise who they contact.

Example

If a sales representative can make a maximum of 10 calls a day, then 10 low value prospects worth £500 each will result in no more than £5,000 in sales for that day.

£500 x 10 calls per day = £5,000 max daily sales

However, if a high-value prospect is worth £10,000, then the best result they can hope for if they can sell to all 10 of them is 20 times more than that — at £100,000.

£10,000 x 10 calls per day = £100,000 max daily sales

In this example, every time a sales rep focuses their time on a £500 opportunity when a £10,000 opportunity exists, they may feel like they're being productive and closing lots of new business, but they're actually far less effective than they could be if they had better focus.

There is a British saying "Don't be a busy fool" — remember that just because you're busy doesn't mean you are busy doing the right things.

The opportunity to improve employee happiness (productivity)

There are lots of benefits to having happy employees, for example, it has been shown to increase productivity between 12–37%. [88]

Happy employees have been shown to be more energised, happier with their jobs and their life, more engaged, and more motivated too.[89] They also take fewer sick days, which is another opportunity cost benefit to your business. [90]

There are lots of things that can impact employee happiness both positively and negatively. For example, if you implement a new tool that employees enjoy working with, you might have a double-whammy benefit of the tool increasing performance and boosting employee productivity due to them being happier.

Employee happiness also increases staff retention, which carries an opportunity cost. For UK companies, the average revenue generated per employee is £118,000. [91] That's the opportunity cost per year of not having an employee. In other words, if the average UK company loses a fee-earning member of staff, they might also lose the ability to generate £118,000 in revenue. In the case of a large business with 100 staff based in London, this figure could be even higher.

Example

Let's see how much a business with 100+ staff could look to save in opportunity costs if it could implement tech that saved each employee 10% of their working day in productivity improvements (a 10% saving on an average eight-hour day equates to 48 minutes saved per day):

Number of employees	100
	X
Estimated revenue per employee	£118,000
	X
Estimated productivity boost	10%
	=
	£1,180,000

In this example, a 10% productivity boost across 100 employees is worth over £1 million per year to the business. This is because that 10% time saved can be allocated to new productive activities, therefore increasing the amount of revenue that the business can generate per employee.

This same approach works for small businesses too, with a 10% improvement in two employees resulting in £23,600 of additional value each year in this example.

The opportunity to reduce bad debts (clients not paying)

One in four UK small businesses suffer from bad debt issues, costing them over £20,000 per year, and bad debts are an issue for companies all around the world. [92]

The average amount of bad debt companies suffer each year is around 1.5% of their revenue, which for large companies is a lot more than £20,000! [93] For large businesses with £100 million in revenue, they could experience £1.5 million in bad debts each year. There's a lot of value in implementing tech ideas to reduce that number!

Fortunately, technology can also help you reduce the amount of bad debts you experience. For example, tools exist to automate reminder emails for payments and online services can automatically chase failed payments for you. If your client can pay more easily, they will be more likely to pay and less likely to become a bad debt.

Test 3: Does your idea reduce errors?

Aside from the litigation example provided earlier, errors are extremely costly for businesses, and making a mistake can have many knock-on effects.

If you're doing something with a risk of error, manually protecting against those errors may be time-consuming and therefore costly. You don't want to spend time fixing quality issues that could have been prevented if the right processes were in place.

Have you ever spent an entire day re-doing work due to someone else's mistake? It's not fun and is extremely demotivating. In the worst case, an error may force you to start again from the beginning — what a waste of valuable time and energy! If you can put processes in place to prevent or eliminate errors, then you're saving money.

Waste caused by errors comes in two forms: direct and indirect. Direct costs are the wasted time and effort to create the error that had to be corrected. If you wrote three pages of a report and had to start again, the direct cost would be your time to create those pages. Indirect costs are the knock-on effects. For example, if the report was a critical part of closing a sale with a new client and re-writing it meant delivering it late or failing to deliver it, then the value of the lost business would be part of the indirect cost.

Errors can also cause significant reputation damage. For example, in the early 1990s, Gerald Ratner famously wiped £500 million from the value of his jewellery business when he said this during a speech: [94]

> *"People say, how can you sell this for such a low price?*
> *I say, because it's total crap!"*
> *– Gerald Ratner*

A company's reputation is intangible, meaning it's hard to assign a value to it. However, this value is very real, and damaging that value can have severe consequences.

A damaged reputation stops new customers from buying and risks losing existing ones. The larger your business becomes, the more exposed it is to reputation-damaging events as the more it has to lose. This reputation risk is especially true since the advent of social media, where messages that damage your brand can be spread rapidly!

To estimate which errors are costing your business, you can work out the types of errors that happen, how many errors of each type are made, and the cost to the business for each error type.

Example

Say you have a graphic design agency creating thousands of leaflet designs for its clients each year, selling each design for £250. If the graphic designer working on the project makes a mistake by misinterpreting the client's brief, they might have to start the job again. This is a waste of £250 to the agency as the designer could have been working on a new client project instead.

If an agency creates 10,000 leaflet designs a year for its clients and makes this mistake with one in five, we can work out the cost to the business:

10,000 tasks x
0.5% failure rate x
£250 cost per mistake

=

£12,500

That's £12,500 worth of errors in the form of reworking leaflet designs.

Catching errors sooner can also save you a lot of time and money. In the software development industry, teams find that the sooner you find and fix an error, the cheaper the resolution. Leave errors until the end to fix and they balloon in size and complexity, sometimes taking more than ten times longer to fix than if caught early.

Test 4: Does your idea generate money?

The easiest way to measure money generated is if your tech idea is a product or enables a service. In this example, you can simply estimate the amount of revenue you hope to make.

But what if your tech idea involves building an internal tool that is not directly responsible for revenue? These ideas do generate money, but indirectly.

For example, if your tech idea does nothing but improve the customer experience, then it may appear that this outcome doesn't generate money. However, happy customers are more likely to buy — so your idea has an indirect benefit to overall company revenue. This improved happiness is even more valuable because selling to new customers is significantly more difficult than selling to existing customers. [95] This means that improving customer satisfaction is a great way to grow your business more efficiently.

Example

Imagine a scenario where you have a great idea for an app that you plan to sell to clients on a subscription basis for £1,200 per year. Is this worthwhile?

To answer this question, we have to consider the cost to develop and maintain the project.

If your annual running costs are £5,000, you can see that the idea costs £15,000 over a three-year period excluding the upfront project cost:

$$£5,000 \text{ } ^{\text{running cost}} \text{ x } 3 \text{ } ^{\text{years}} = £15,000 \text{ } ^{\text{3-year cost to run}}$$

In addition to these running costs, you also need to pay upfront costs to build the app, so we'll factor in £20,000 for that, adding this to our three-year cost:

$$£15,000 \text{ } ^{\text{3-yr running cost}} + £20,000 \text{ } ^{\text{upfront costs}}$$
$$= £35,000 \text{ } ^{\text{total 3-yr costs}}$$

This gives us a total three-year project cost of £35,000. This might seem like a lot of money, especially if you're a new start-up business with limited cash, but there's more to this project than how much it costs to build and maintain.

To decide whether this project is a worthwhile investment, we need to figure out how much money it might generate for us.

Each new customer we sign up to a subscription makes £1,200 for the business (£100 per month x 12 = £1,200).

To figure out the total value of this project, we need to decide how many customers we believe will sign up to pay each month.

As this new project has no marketing budget, let's go with one 1 new customer each month.

In month 1, we make £100 for the single customer, but in month 2, we have two customers on board. So in the second month, we make £200, and £300 the next, and so on:

Monthly revenue per customer:	£100		

Month:	Jan	Feb	March
Number of paying customers:	1	2	3
Monthly revenue:	£100	£200	£300

By the end of month 3, we've made £600 in total since the launch.

$$£100^{month1} + £200^{month2} + £300^{month3} = £600$$

Let's extend this revenue estimate for a full 12 months:

Monthly revenue per customer:	£100											
Month:	Jan	Feb	Mar	Apr	May	Jun	Jul	Aug	Sept	Oct	Nov	Dec
Number of paying customers:	1	2	3	4	5	6	7	8	9	10	11	12
Monthly revenue:	£100	£200	£300	£400	£500	£600	£700	£800	£900	£1,000	£1,100	£1,200

Total annual revenue:	£7,800

If we acquire and keep one customer each month, we stand to make £7,800 in the first 12 months from launch.

$$£100^{month1} + £200^{month2} + ... + £1,100^{month11} + £1,200^{month12}$$
$$= £7,800$$

But to get to the amount of revenue we'll make over three years, we can't just multiply this number by three (which would give us £23,400 revenue over three years). This is because at the start of the second year, we don't start from zero customers but 12.

If we extended this graph for a full three years, then the total revenue would be **£66,000**:

$$£100^{month1} + £200^{month2} + ... + £3.5k^{month35} + £3.6k^{month36}$$
$$= £66.6k$$

Remember we worked out that the total three-year project cost would be £35k? [96] Is it worth spending £35k for the chance to make £66k over three years? We double our money, so quite possibly. Your answer will depend on your risk tolerance and whether you're willing to wait three years to achieve this figure.

Now remember we picked a conservative one-customer-per-month signup rate. What might happen if we were able to acquire five sign-ups each month?

Monthly revenue per customer:	£100

	Year 1											
Month:	Jan	Feb	Mar	Apr	May	Jun	Jul	Aug	Sept	Oct	Nov	Dec
Number of paying customers:	1	6	11	16	21	26	31	36	41	46	51	56
Monthly revenue:	£0.1k	£0.6k	£1.1k	£1.6k	£2.1k	£2.6k	£3.1k	£3.6k	£4.1k	£4.6k	£5.1k	£5.6k

	Year 2											
Month:	Jan	Feb	Mar	Apr	May	Jun	Jul	Aug	Sept	Oct	Nov	Dec
Number of paying customers:	61	66	71	76	81	86	91	96	101	106	111	116
Monthly revenue:	£6.1k	£6.6k	£7.1k	£7.6k	£8.1k	£8.6k	£9.1k	£9.6k	£10.1k	£10.6k	£11.1k	£11.6k

	Year 3											
Month:	Jan	Feb	Mar	Apr	May	Jun	Jul	Aug	Sept	Oct	Nov	Dec
Number of paying customers:	121	126	131	136	141	146	151	156	161	166	171	176
Monthly revenue:	£12.1k	£12.6k	£13.1k	£13.6k	£14.1k	£14.6k	£15.1k	£15.6k	£16.1k	£16.6k	£17.1k	£17.6k

Year 1 revenue	£34.2k
Year 2 revenue	£106.2k
Year 3 revenue	£178.2k
Total 3 year revenue:	£318.6k

$$£100^{month1} + £200^{month2} + ... + £17.1k^{month35} + £17.6k^{month36}$$
$$= £318.6k$$

In this example, £318.6k isn't far off a 10x return on your initial £35k investment. It's starting to look a lot more attractive!

This example illustrates why your sales and marketing approach matters so much to your tech idea — as the difference between closing one and five new customers per month is huge.

I appreciate that drawing out tables to work out your monthly revenue over three years is useful, but certainly not something you can do in 60 seconds on the back of an envelope!

Fortunately, there's a handy trick to estimate these numbers so you don't have to draw the full table each time.

First multiply the number of customers you expect to acquire each month by the number of months, then by the number of months again.

For five customers per month over 36 months, this gives us a total of 6,480:

$$5^{customers\ per\ month} \times 36^{months} \times 36^{months} = 6,480$$

And then half it:

$$6,480 / 2 = 3,240^{payments}$$

> The 3,240 figure is an estimate of the absolute total number of monthly payments made by all customers over three years. So now, all we have to do is multiply this number by our monthly fee to get a revenue estimate:
>
> $$3{,}240^{\text{payments}} \times £100 = £324{,}000$$
>
> £324k is a very close estimate to the £318k figure we arrived at by drawing a table, close enough for a back-of-envelope calculation!

Another way to generate money is if existing customers stay for longer, which is why most companies measure customers acquired *and* lost, then try to improve both those factors. The loss rate for existing customers is called the "attrition rate".

Retaining existing clients for longer periods of time is a huge business opportunity, and relatively small increases in customer retention can have a surprisingly large impact on revenues and profits. [97]

This is because acquiring new customers is a difficult and expensive activity. You've got to identify who they are and how to reach them, and direct a significant number of resources (time or money) to convince them of your value so they buy from you.

By comparison, existing customers are already convinced of your value as they've purchased from you already.

Example

If the average customer who signs up to buy your product or service spends £1,000 a year and only stays a customer for a year on average, then the lifetime value of that customer[98] to your business is £1,000 (£1,000 x 1 = £1,000).

If you manage to sell your product to ten new customers in a year, you stand to make £10,000 in one year from the acquisition of those customers with this attrition rate.

Average revenue per customer:	£1,000

	1
Year:	1
Number of customers aquired:	10
Number of active customers:	10

Revenue for the year:	£10,000

Or put more simply:

$$10^{\text{aquired customers}} \times £1{,}000^{\text{avg lifetime value}} = £10{,}000$$

If you maintain this level of customer acquisition and customer attrition for four years, each year you acquire ten customers but also lose ten customers. In practice, this means your revenue each year stays at a constant £10,000 figure as the number of active customers stays the same, with acquisitions perfectly matching attrition:

Average revenue per customer:	£1,000
Attrition rate	100%

Year:	1	2	3	4
Number of customers aquired:	10	10	10	10
Number of customers lost	0	-10	-10	-10
Number of active customers:	10	10	10	10
Revenue for the year:	£10,000	£10,000	£10,000	£10,000

Total 4-year revenue	£40,000

In this example, we make £40,000 over the course of four years from our efforts to win 40 customers over this period.

Year:	1	2	3	4
	...	...	...	...
Revenue for the year:	£10,000	£10,000	£10,000	£10,000

Total 4-year revenue	£40,000

Or

$$40^{\text{aquired customers}} \times £1{,}000^{\text{lifetime value}} = £40{,}000$$

As we lose 100% of our active customers in a year, our attrition rate is 100%.

What if we could implement an innovative feature or change of approach that halves this attrition rate? Rather than losing all of our active customers each year, requiring new ones the following year, we only lose half of them each year.

A 50% attrition rate means that if we acquired ten customers in year one, we'd lose five of them at the end of the year, leaving five active users.

This new lower attrition rate means that in year 2, we have 15 active customers rather than ten. If we follow this acquisition and attrition rate for four years, we get the following scenario:

Average revenue per customer:	£1,000			
Attrition rate	50%			

Year:	1	2	3	4
Number of customers aquired:	10	10	10	10
Number of customers lost	0	-5	-7.5	-8.75
Number of active customers:	10	15	17.5	18.75

Revenue for the year:	£10,000	£15,000	£17,500	£18,750

Total 4-year revenue	£61,250

By halving the attrition rate, we've grown the four-year revenue from £40k to £61k, which is over 50% more revenue from the same 40 acquired customers. In this example, halving the attrition rate is worth £21k to the business.

What if we could drop the attrition rate to just 25% rather than 50%?

Average revenue per customer:	£1,000			
Attrition rate	25%			

Year:	1	2	3	4
Number of customers aquired:	10	10	10	10
Number of customers lost	0	-2.5	-4.375	-5.78125
Number of active customers:	10	17.5	23.125	27.3438

Revenue for the year:	£10,000	£17,500	£23,125	£27,344

Total 4-year revenue	£77,969

With a 25% attrition rate, these 40 acquired customers are now worth £78k rather than the original £40k — approximately double the revenue!

Dropping our attrition rate from 100% to 25% is worth £38k extra in this example.

$$£78k^{25\% \text{ attrition revenue}} - £40k^{100\% \text{ attrition revenue}} = £38k$$

Now imagine the impact that an attrition-improving tech idea could have for a business making hundreds of thousands or millions each year. These tech innovations can deliver huge value to businesses smart enough to pursue them.

Finally, you can generate money if existing customers chose to pay more to upgrade their existing product or service with you — a technique called up-selling. It's somewhere between 50% and 80% easier to sell to existing customers compared with new prospects, so this is a powerful technique that can't be ignored. [99]

Sometimes, new tech ideas can be focused around creating new products or services, or improved tiers of service, that encourage customers to spend more in exchange for additional features.

If you have an existing suite of products, consider whether your idea enables you to sell to existing users, determine the value they might place on it, and decide what they're willing to pay for it.

Example

Picture a business with 500 customers paying £1,000 per year for their tech service.

What if we could develop a brand-new app that greatly improves the quality of the service? A new feature could add value to end customers who upgrade to this attractive new offering.

> If your changes deliver enough benefits to the customer, you might double what you make from each client who upgrades — from £1,000 to £2,000.
>
> Let's work out what this new set of features is worth to your business if we can convince one-in-five (20%) customers to upgrade:
>
> $$500^{\text{existing customers}} \text{ x } 20\%^{\text{upgrade rate}} = 100^{\text{upgraded customers}}$$
>
> As our higher price package costs £1,000 more than the old one, this new set of features is worth £100,000 in additional annual revenue!
>
> $$100^{\text{upgraded customers}} \text{ x } £1,000^{\text{extra per customer}} = £100,000^{\text{extra revenue}}$$

You should review and combine these methods of assigning a value to your idea in terms of the revenue it makes. Once you have a figure, compare it to the upfront and ongoing running costs to create and support your tech idea, then decide whether your idea is worthwhile.

Test 5: Does your idea add value as a brand or intellectual property (IP) asset?

A brand is the impression that people have about a product, service, or company. It's an idea that's been spread, a gut feeling guiding people's expectations when they buy.

If it's instilled into enough people, this set of brand beliefs carry value – some companies like Coca Cola have a brand value that on its own is worth billions. [100]

But why are brands like Coca Cola worth so much?

One reason is identity — when you think of Coca Cola, it evokes a certain feeling or place. [101] Maybe you have positive memories of drinking Coke on holiday or in a social setting. You also know if you buy the drink, the experience will be the same or similar to your past enjoyable experiences.

All of these thoughts, feelings, and memories combine so when you're at a shop or restaurant about to buy a drink, it's the Coca Cola brand you reach for. You don't even think about it; you buy out of habit.

This habitual tendency results in recurring business for Coca Cola, with consumers buying their products every day. These habitual purchases are worth big bucks for global brand known and loved by billions.

Another example is Apple Computers, whose brand alone is worth hundreds of billions of dollars. [102] I'd argue that Apple's business is so valuable because of its brand — they are a brand that happens to be a tech company, more than a tech company that has a valuable brand. Apple sell technology, of course, but do you think their user base is genuinely concerned with the technical specs of their products above all else?

Apple uses its brand in the same way as major designer brands such as Gucci and Armani — people are buying status and what it means to associate themselves with that brand.

This powerful, status-driven brand image allows Apple to command a high price compared to competitors selling similar products — this high price says something about quality and prestige. People know that Apple products cost more than other brands, and that knowledge inflates the status of those who buy their products.

If building your tech brand helps you to sell more over time or command a higher price, that value can be measured. Or even better, your idea might be so strong that it causes others to shout about your brand and recommend it to others, creating a viral feedback loop.

But how can we put a cash value on the brand values that your tech idea aims to support? We can compare what customers are willing to pay with a strong brand vs. no brand, and see what impact that has on their lifetime value. [103]

If you have an existing business, you can work out the existing average customer lifetime value. For new start-ups, you will need to estimate or use industry averages.

Example

Imagine our product is a £400 smartphone with little to no brand value.

On average, people upgrade their devices once every two years. We'll assume that because your brand is weak and the competition is fierce, it's unlikely that they'll repeat-purchase from you.

This means our lifetime value is the value of the device: £400, which works out at £200 of value per customer per year, for a maximum of two years:

$$\text{No brand lifetime value} = £400$$

But what if we did have a strong brand, one as strong as Apple? We could find that customers repeat-buy for a decade or more. If we could extend their loyalty to a decade, this would mean that customers buy five times over their life, four more times than if our brand was poor:

$$£400^{\text{sale price}} \times 4^{\text{additional lifetime purchases}}$$
$$= £1,600^{\text{Additional value per customer}}$$

£1,600 additional lifetime value per customer is huge — if you had just 1,000 customers, then the brand brings £1.6 million in additional value.

But the brand value doesn't stop there as a strong brand could enable you to extend the purchase frequency and the average sale price:

$$(£1,000^{\text{higher sale price}} \times 5^{\text{new lifetime sales}}) - £400^{\text{no brand lifetime value}}$$
$$= £4,600^{\text{Additional value per customer}}$$

In this example, having a strong brand makes people spend more and buy more, a double whammy effect that increases the average customer lifetime value by over 10 times! If you had 1,000 customers, the brand value is £4.6 million:

$$£4,600^{\text{Additional value per customer}} \times 1,000^{\text{customers}} = £4.6 \text{ million}$$

The value of your brand can grow your business success significantly, allowing you to charge more and sell more frequently to your existing customers.

Unbelievably, we are still not done yet; there's another factor that influences the value that your brand brings: recommendation value. If your brand values are strong enough, your existing customers will recommend your business to their friends, family, and colleagues.

Example

In our last example, our brand was worth £4,600 in additional value, which was £5,000 in total value if we add in what our unbranded business would be worth.

If 1 in 5 customers recommends someone who signs up, we have a recommendation multiplier of 20% (1 / 5 = 20%). If we only go one level deep, this gives us the value of referral business:

$$\text{£5,000}^{\text{branded lifetime value}} \times 20\%^{\text{Referral sales}}$$
$$= \text{£1,000}^{\text{value of referral business}}$$

It's actually even more if we go several levels deep, as each £1,000 also generates 20% in referral sales, adding £200. Then each £200 ads a further 20%, adding another £40.

If we follow this all the way down, the total value of the referral business in this example is about £1,248.

Let's add this value of the referral business to our total:

$$\text{£4,600} + \text{£1,248} = \text{£5,588}$$

That's nearly 14x more revenue than in our first example where we only made £400 from each customer:

$$\text{£5,588} / \text{£400} = 13.97 \text{ times}$$

The lesson? Ignore brand value at your peril! This is also why it's usually much better to compete on brand rather that price.

Test 6: Does your idea boost the value of your company (profit multiplier)?

There is no one-size-fits all rule for how to value a company. The price that companies are bought and sold for depends on many factors, including the sector it operates in, the capabilities of the management team, the unique competitive advantage it has, the size of the business, the brand, the Intellectual Property (IP) it holds, its dominance in the market, the amount of cash or other assets it holds, and the list goes on!

But what if I told you there are ways to grow the value of your company without having to grow your revenue or profitability?

One way that acquisitions place a value on a business is by looking at the profit it makes and applying a multiplier. This makes sense because what someone is willing to pay to buy a company is influenced by what that company stands to make them in returns.

Example

Imagine I offered you the opportunity to buy "company X" for £10,000 — is that a good deal? If this is the only information you have, it's impossible to know, so your answer will rightly be "It depends".

Now what if I told you this company consistently makes £5,000 in profit each year, meaning you'd have your money back in two years' time and stand to make £5,000 in returns every year after that?

That sounds like a fantastic deal — you get your money back in just two years, and from year 3 onwards, you are winning!

$$£10,000^{purchase\ price} / £5,000^{annual\ returns}$$
$$= 2\ years'\ ROI$$

In this example, you are happy to buy a company for a 2x profit multiplier because the price you paid is double the profit.

Now what if the company made less, £3,333 per year — [104] are you still happy to pay £10,000 for it?

> Personally, I'd be happy with that level of return, and if you are also happy with [105] those numbers then this shows you'd be prepared to buy a business with a 3x profit multiplier.
>
> But how about a 4x profit multiplier — would you be happy with that? And what about a 10x or 20x multiplier?

As the profit multipliers get higher, you probably want to know more information about the company before deciding if you'd be happy to buy it.

For example, to justify purchasing at a higher multiple, you may want to know how stable the business is, or what sector it operates in, or whether it has a good chance of growing in future years. You may be happy to pay £10,000 for a company that currently makes only £100 per year (a 100x multiplier) if in three years' time you are confident that it has the potential to make £5,000 per year, giving you a 2x future profit multiplier.

Companies and investors go through a similar thought process when deciding on what to pay for a company. This means if you can give them confidence in the company's prospects, you may command a higher profit multiplier and therefore a higher overall company value as a result.

Companies invest in technology to try to grow their profit multiplier and make their business more valuable relative to their profits. They invest in tech as a strategy to achieve value growth for several reasons:

1. Technology is seen as the future, and investors may feel that non-tech-enabled companies are likely to be out-competed by those that are. This can lower non-tech-enabled valuations compared to companies making good use of tech.

2. When a business implements tech, this often forces them to have robust and scalable processes. Such processes, supported by the tech, can give companies the foundations to grow as they are well-defined and often automated. Tech processes scale well from an operational perspective as they run on computers, which tend to be cheaper than employing people to conduct the processes.

3. Technology can scale. It may be possible for a tech-enabled company to grow globally in ways that a non-tech, local company cannot. If a company can become tech-enabled, this allows them to create products and services with international reach. This makes a company's growth prospects more attractive and can have upwards pressure on what investors are willing to pay as a multiplier of profits.

If you're looking to raise rounds of external investment or sell your business in the future (called an exit), then taking action to execute your tech idea could be the most powerful tool available to protect and grow the value of your business.

Example

Imagine we have a company that makes £100k each year in profit. This company isn't a tech-enabled business, so let's give them a profit multiplier of just 5x.

This gives us a current company value of £500k:

$$\text{£100k}^{\text{profit}} \times 5^{\text{profit multiplier}}$$
$$= \text{£500k}^{\text{company valuation}}$$

Now what if investments in tech helped to grow their profit multiplier from five to ten? This grows the company valuation to £1 million:

$$\text{£100k}^{\text{profit}} \times 10^{\text{profit multiplier}}$$
$$= \text{£1 million}^{\text{company valuation}}$$

This means tech enablement is worth £500k to the business:

$$\text{£500k}^{\text{old value}} - \text{£1million}^{\text{new value}}$$
$$= \text{£500k}^{\text{ total additional value}}$$

Consider the profit your company makes, review industry profit multipliers, and compare them to the tech.[106] That way, you'll get an idea of how much you could increase your profit by if you became tech-enabled. Use the difference in the multipliers to estimate how much your company could be worth with the right tech strategy in place.

Test 7: Does your idea address a large enough market?

For product-based businesses, the total addressable market is the total available opportunity to make revenue from a product or service in a certain market. For internal tools, the maximum opportunity size is a ceiling on what the idea could be worth to your business when fully implemented.

In the last chapter, we worked out different methods to put a value on your idea, but we didn't consider ways to limit the maximum possible size of the opportunity. Opportunity size is an important consideration because even if you could double your business each year, there is a limit on how far this pace of growth can take you for some tech ideas.

Think of it like a big fish in a small pond — you can grow your influence but at some point you'll be limited by the size of the pond. In other words, it's all well and good sitting back and looking at your calculations of a multi-million-dollar revenue business, but in reality, your idea will occupy a market, and that market has a size. It's no good having million-dollar revenue forecasts if the total market size is far smaller than that.

The goal of a market size test is to judge the potential impact of your idea in financial terms alone. If your idea generates money, then how much could it generate in a perfect world where you take the whole market? If your idea saves money, then what is the most it could save within the market it will be applied to?

You may be wondering how you can figure out the market size without conducting in-depth market research or buying an expensive report. But that's not the point of the exercise. You should be able to estimate all of these things with nothing but your mind and a few Google searches.

Example

Dave's tech idea was to automate the process of recording videos at university graduation ceremonies. He wanted to provide software for the videographer to use, allowing them to easily split the footage based on the names announced.

Dave's business model was to sell his tech to the videographer, taking a small cut of the videos sold to graduates. The goal was to save the videographer time in processing each graduation and make a percentage from the sale of each video. If a videographer spent 10 minutes editing each video, the tool could save them a lot of time, hence the idea's value metric was time-saving.

I asked Dave about the market size, and a quick Google search showed that 500,000 students entered higher education last year. Being optimistic that all of them go on to graduate, 500k seems like a big number, and a good start.

Dave expected each videographer to sell to 10% of graduates (50k of 500k), and to charge a 10% commission on whatever the videographer sold.

If we assume that each buyer spends £20 on average for their video, the calculation is pretty simple:

$$£20^{\text{video sale}} \times 10\%^{\text{commission}} \times 50{,}000^{\text{students that buy}}$$
$$= £100{,}000 \text{ annual revenue opportunity}$$

That's a maximum £100k revenue each year for the whole UK market, of which Dave would probably take only a small portion.

I felt terrible as I saw the colour drain from his face when I pointed this out. Fortunately, he was yet to spend too much money on the idea.

The small addressable market doesn't mean that Dave's idea is dead, but his current marketing and strategy certainly is. Maybe he could repair the situation by finding a larger addressable market for the same tech idea and start by specialising in university events before broadening his target market.

Don't make the same mistake as Dave. Take some time to consider the size of your market. One option is to buy industry reports aimed at specific markets, as these will often estimate the potential market size.

Another option is to research your competition. If a handful of large businesses make up most of the market size, you can look at the annual reports on their website. These reports contain lots of financial information, including where they get their revenue from and sometimes how many customers they have. You can then combine the revenue or customer data from a handful to get an estimate of the market size.

Finally, you can speculate on the market size by making assumptions based on different demographic data. For example, if your product or service is aimed at certain kinds of working parents, you can search for how many working parents there are in your country.[107] Let's say there are 10 million working parents and you believe that one in five have a certain set of interests, you might estimate the total addressable market as 2 million people.

For internal tools or elements of your project that add value but don't generate revenue, you can look at other metrics to estimate the total opportunity value.

Let's say you're building an internal tool (not a revenue-making product) that will save lot of administrative time for your team proportional to the number of customers. There are two main factors that influence the opportunity value: the number of staff and the number of customers.

Now estimate how many customers you hope to have and the size of your team in five years' time, then use the techniques from the last chapter to calculate the maximum opportunity value if you grow the tool to this size.

Understanding the opportunity size influences the amount you're willing to invest to implement and grow your tech idea. Ideas that have a lot of growth potential can be more likely to achieve large cash investments or get buy-in from senior people inside your business.

For internal tools, you can measure the opportunity size by looking at the maximum value your idea could bring to your business in the future, such as five years from now.

Bonus tests: Indirect ways to assign a value to your idea

Here are some honourable mentions of other methods you can use to figure out the value your idea brings. As with the previous examples, try to think of ways you can put a number on the features and benefits of your tech idea.

Your tech idea can deliver value if it improves:

1. **Product quality:**

   If something feels like a higher quality, people might pay more for it. Lots of factors contribute to quality — with a pair of shoes, it might be the fabric or the stitching used. For an app, it could be the colour scheme and animation effects.

2. **Convenience and user experience (also called UX):**

   People like their lives to be made easier. If you can simplify a process and make it consume less brain power, then you deliver more value to the user.

3. **Business data, insights, and information (make better decisions):**

   If you have a process that doesn't involve an app and you implement an app for the first time, chances are that you'll end up collecting a ton of business data that you didn't have before in addition to the core value the app brings. Data can influence business decision-making in ways you never realised possible, changing the way you act and delivering improved results.

4. **Speed (of delivery, service, or information):**

   When you replace manual tasks with software, you increase the speed that things can be done. Would you rather wait until the weekend, drive into town, spend an hour looking around the shops, then buy the product you're looking for and drive home... or go online for 10 minutes, find what you want, and order it for next-day delivery for free? Speed of delivery can add tremendous value.

5. **Enable access:**

   Your solution might enable access to your services in ways not possible before, in effect widening the available market for what you offer. Take a look at how internet banking helped those living in remote rural communities: it's now possible for people to bank from the comfort of their own home, vastly increasing the number of options available to the end consumer, as well as the reach of banks who leverage this technology to reach those customers.

6. **Goal achievement:**

   By nature, people have goals because they value what they're trying to achieve, and these goals don't necessarily have to be commercial. Take a look at Olympic athletes — often they don't care about the money and are in it for the satisfaction of achieving their goals. Develop something that supports these goals and you'll deliver significant value, especially if you help people achieve something that they previously thought impossible.

Closing words

Now you know all the different ways you can assign a value to your idea, shortlist one or more you think apply to you and perform a back-of-envelope test to estimate the potential value.

The value your idea brings should be higher than the time and cost to build and maintain it (we'll cover how to implement your idea later in Part 3 of the book). If you find that the potential value of your idea is much more than the cost to implement, this ratio of value to cost will give confidence that it's well worth the investment.

Don't worry if you perform your back-of-envelope tests to find that your tech idea doesn't add much value and makes you question yourself. Realising this early is a good thing as it's easy to change your approach now — called a pivot — rather than later in the process. Go back to the techniques in earlier chapters to modify your idea or the problem you've chosen to solve.

Key takeaways:

1. One way to assign a value to your idea is to identify how much revenue it could make over its life.

2. Time is money, so if your idea can save time, it may add a lot of value to your business.

3. Lost opportunities can be extremely expensive. Identify and capitalise on opportunity cost to boost the amount of value your idea can add.

4. It is commonplace for businesses to be valued as a multiplier of their profits, and you may be able to grow the value of your business by implementing strong tech-enabled processes.

5. Compare the value that you estimate your tech idea will brings to the cost to implement it.

6. The total addressable market is the total available opportunity to make revenue from a product or service in a certain market.

7. You may have a great tech idea, but if the total opportunity size for that idea is too small, the cost and risk implementing the idea may not be worth the return.

8. If your idea fails all the value tests, then pivot and see whether you can approach the problem in a different way, to a different market, or even find a different idea altogether.

CHAPTER 5:
WHY YOUR VISION MATTERS

"Create the highest grandest vision possible for your life, because you become what you believe."
— Oprah Winfrey

Chapter Relevance	
New Internal Tool	**New Product**
→	↑

This chapter explains why you need a vision that influences how you set your goals. While it's useful to quantify the value your idea creates, like we did in the last chapter, if you lack vision then you risk getting lost and wasting precious time and money.

As you shape your ideas over time, you will realise that there are many competing ways you can choose to add value. How should you decide which values to develop over others? Vision is a powerful tool that will help you to pick the correct 'right ways'.

Let me give you an example of how businesses define their vision. Here is Google's vision:

"To organize the world's information and make it universally accessible and useful."

And Microsoft's:

"We believe in what people make possible. Our mission is to empower every person and every organization on the planet to achieve more."

Notice how both of these visions are narrow in some ways, and broad in others? They specify a high-level objective but the objective remains stable in an ever-changing world. These vision statements will make sense 20 years from now and did 20 years ago. Most Fortune 500 companies have a vision or mission statement, though not every small business does.

But why does having a vision matter? Let me answer that with another question: How do you plan on achieving your next goal once your current goal is met? Do you plan on celebrating the achievement before you close down your business, job done, right? I doubt it.

If you have a goal to achieve or do something, then when you achieve that thing, the game is over. Games that end once we achieve something are called finite games.

Vision is important because business is an infinite game. [108]

Infinite games go on forever — the rules can change, the players can change, and the goals can change. Players of an infinite game either keep playing or drop out, and to 'lose' is to stop playing. A finite board game would be to play and win, then you're done. An infinite game would be having the vision to be the best board game player you can be for as long as possible. [109]

You need a vision to guide you from goal to goal, objective to objective, and task to task in the infinite game that is business.

Unless you plan on ending your business once you meet your goals, business is an infinite game, and hitting specific business targets and goals are finite ones. You may need to hit your finite targets to stay alive in the infinite game of business but you risk complete failure if you pursue finite goals that conflict with your infinite game vision.

It's easy to fall into the trap of focusing too heavily on the short term, like how much money you can make today. But if you only make decisions based on short-term outcomes, you won't set a long-term vision or goals. Such short-termism can be detrimental to your organisation, confuse your customers and staff about what you stand for, and make it difficult for people to know what kind of business it is.

Aiming to achieve certain monetary targets is also a finite game. Focusing on money alone might mean you fatally miss a move in the game, one you may have noticed if you were playing a visionary infinite game. You may achieve short-term success but fail in the long term.

If you come up with a solid idea that passes your back-of-envelope value tests, then take five minutes now to think about whether the strategy you have in mind aligns with your vision and company mission.

Small ideas might be a low-risk safe bet — if a company does something obvious like eliminate a paper-based process thanks to a digital one, this most likely aligns with the goals of every business on the planet. However, if your mission is to provide the best in-person customer service at a premium cost, then automations that lower the cost but also the customer experience are detrimental to the business vision, making you risk losing the infinite game.

When I was brand new to business, I honestly thought that the idea of having a vision was nonsense, stuff that motivational speakers told you to do because it sounded good. Because having a vision is, of course, visionary, and people want to come across as visionaries.

When my business was just me, I could do things I wanted to do at that moment and set a strategy based on my current goals. I didn't see the need to define a vision. Yet, as my team grew and I started to employ people in management roles, I had an 'Aha!' moment. I realised that as you grow a business, you also grow the responsibilities required from your team. If you have a small team of up to five people, it's possible to simply define a strategy, delegate it, and check in to ensure that what you wanted to happen is happening.

However, as my team grew, I realised that effective delegation isn't just about allocating tasks to my team; it's also about delegating strategic decision-making, which includes empowering them to autonomously deal with situations that don't fit the current plans. But how can your team make strategic decisions on their own if they don't have a shared vision about what the company is trying to achieve?

Sure, they can make decisions, but how will they know the right ones without the guiding force of your vision? And if they make a mistake and implement a poor strategy, is it their fault or yours for not giving them enough guidance?

My vision, and the vision of my company, is that I see a future where technology continues to enrich people's lives and grows business productivity. I want to ensure that this kind of future will happen. I want to achieve this future by supporting or enabling businesses of all sizes to embrace technology to innovate, especially those that don't think of themselves as tech businesses, because I see a future where every business is a tech business, and I think we will be there sooner than you think.

As you continue to read, I hope you can see how this book is also part of that vision.

Closing words

Take a moment to consider the vision driving your actions — it shouldn't take long. Think about what you want to achieve, which could be a specific goal or target. Now ask yourself why you want to do that, and consider the answer. Why is it important to you?

You can keep asking 'why' questions until you get to the bottom of it. If done correctly, you will find one or maybe two high-level objectives driving your business decisions. Now consider what advice you would give to ensure that someone always makes decisions that support this high-level objective. Be sure to explain both the 'what' and the 'why'. Now write it down because *this* is your vision.

Key takeaways:

1. You should have a vision and mission that you and others in your team can use to guide all of your strategic decisions.

2. Although you may play a finite game to win in the short term, the most successful businesses get good at playing the infinite game, and you should too.

3. Consider your goals and ask yourself a series of 'why' questions to determine the high-level reasons you have those goals. Explain the 'what' and 'why' to arrive at your vision.

PART 2: IMPLEMENT YOUR IDEA

Establish your value, prioritise your idea,
plan the best approach, and begin your build!

CHAPTER 6:
DEFINE YOUR VALUE PROPOSITION

"Some people say, "Give the customers what they want." But that's not my approach. Our job is to figure out what they're going to want before they do. I think Henry Ford once said, "If I'd asked customers what they wanted, they would have told me, 'A faster horse!'" People don't know what they want until you show it to them. That's why I never rely on market research. Our task is to read things that are not yet on the page."
— Steve Jobs, former CEO of Apple

Chapter Relevance	
New Internal Tool	**New Product**
↑	↑

This chapter will explain what your value proposition is, why it's important, and how you can use it to prioritise the execution of your idea in the most effective way.

Every idea you have will come with a cost, as doing one thing now delays something else until later or perhaps not at all.

Reflecting on the previous chapters, your idea must add value to stand a chance of being successful in a way that aligns with your vision. The more value you add, the higher the chance of success becomes.

In short, your value proposition is a statement of the value that your product or service brings to another person — usually the customer or end user. Products have value propositions and businesses have value propositions. In fact, even *you* have a value proposition.

Think of the last time that you were hired to do a job. The role and pay offered were because of the value you brought to that company. If you didn't add enough value in the areas the employer required most, then you wouldn't have been offered the job. But if you understood and aligned what you said to the employer's needs, it would have increased your chances of being hired. The power of value propositions is demonstrated everywhere — whether you notice them or not!

That's why every business and product or service within a business should have its own value proposition, designed to focus your attention and your business efforts. Among other things, value propositions bring clarity to your marketing. You must be clear on who your target customer is to communicate and sell most effectively to them.

For example, here is the value proposition of my development company: [110]

> "We can deliver your innovative, technically complex project, using the latest web and mobile application development technologies.
>
> Scorchsoft develops online portals, applications, progressive web apps, and mobile app projects. With over a decade of experience working with hundreds of small, medium, and large enterprises in a diverse range of sectors, we'd love to discover how we can apply our expertise to your project."

But why this value proposition?

If you're looking to build an app or software project, you may recognise that your tech idea has lots of features and complexities. You know that the approach you take is critical to your success, and if you make the wrong choices, you risk a poor-quality product, missed business objectives, or costly rework time.

My company's value proposition addresses these needs, recognises clients' common interests, and explains how we can help. It then gives examples to build trust and demonstrate that we can deliver on the promises outlined in the value proposition.

Your value proposition should also follow a similar format. You should try to do the following:

1. Explain what you offer in a compelling and concise way.

2. Be clear what makes you different (your unique selling point).

3. Back up your claims in a way that builds trust.

4. Use keywords or phrases relatable to your target audience.

5. Include a hook or a call to action (optional).

Here's another example of a powerful value proposition from project management software provider Atlassian. This is their value proposition for one of their most popular tools, Jira (which can also be considered a 'tech idea'):

> *The #1 software development tool used by agile teams. The best software teams ship early and often. Jira Software is built for every member of your software team to plan, track, and release great software. Trusted by over 50,000 customers worldwide."* [iii]

Can you spot the offer, unique selling point, trust factors, and key terms in this value proposition?

Atlassian makes a bold statement about how good they are using the word "agile", which they know their audience in the software development industry will immediately relate to.

Atlassian tells the user what "the best software" teams do, which is a subtle hint that if the user considers themselves one of the best, they should use these tools too. They spell-out what the Jira software is and does. Finally, they back up their claim of being #1 by stating the number of clients who use their software.

Now let's look at the social network Twitter:

> *Follow your interests. Hear what people are talking about. Join the conversation. See what's happening in the world right now.*

Notice how they talk about the benefits of using Twitter, not just the features of the platform? They could easily have written something dull such as "Post 200 and something character messages to your public profile, follow people, and be followed", but that isn't the reason why people use the service. That's why Twitter chose to focus on why people post and why people read other people's posts. It's about finding and engaging with interesting, relevant, real-world, real-time micro-news and updates.

Value propositions like these focus the attention of the customer, but also the stakeholders in your business.

One challenge, however, is that it's common for businesses to engage a wide range of potential clients in different niche categories and with different interests. This is when value propositions become a little trickier, as it's difficult to write a single message that everyone in your audience relates to.

For example, think of someone who buys a car with a large engine (or powerful electric motor). A big engine means more power and a faster car.

One type of customer values the car for the experience of driving fast on winding country roads — they like the thrill of driving. Another type of customer values the car because the high torque power enables them to hook up their caravan or trailer and still be able to drive at a reasonable speed uphill. The feature of the product — a big engine — is the same, but the benefits are very different.

If you approached the fast-winding-roads customer with the promise of being able to tow a heavy caravan, you'd misalign your value messaging. Poorly aligned messaging like this means you probably won't sell the car to them, even though the product you're selling aligns perfectly with their interests.

Let's reflect on the value proposition of another kind of business — a commercial gym. Gyms are an interesting example as there are a wide range of people from different backgrounds who are members for different reasons.

Some people join the gym because they want to lose weight, others want to get in shape after having a baby, some want to compete in weightlifting or powerlifting, while others are interested in bodybuilding, and some attend for the gym's social culture.

You can see why it may be sensible to approach each of these audiences with a slightly different value proposition message that appeals to them personally. Let's have a go at writing one for each of them:

If targeting powerlifters and those interests in elite sports performance:

"Max-Power Gym has 12 squat racks and over 50,000 Kg of calibrated Eleiko and Rogue plates, enabling you to achieve your full potential without having to wait for equipment. Our team of experienced strength coaches can help you achieve your goals and have over 50 years of combined powerlifting experience."

If targeting recent mothers:

"Friendly Gym runs regular classes every Monday and Friday to help recent mums work on their tums. With over 50 five-star reviews, our friendly team of instructors will help you to ease yourself back into shape, keeping you motivated by connecting you with other new parents doing the same."

If targeting a generic audience of people looking to lose weight:

"Fat-Beater gym will help you to shift those Christmas pounds. Our 12-week program has helped hundreds of people just like you, and includes full membership of our gym and facilities, as well as a dedicated coach to help you on your weight loss journey."

All three of these value propositions are selling the same services: access to gym and a personal trainer or coach, but the language and messaging used to promote those services to each audience are very different.

These differing value messages illustrate why you need to think about your different customer types, called personas, before you write your value proposition. This is why it's common for businesses to have several value propositions for the exact same product depending upon which customer you're speaking or marketing to at the time, plus one broader and less niche value proposition that all personas will relate to.

With these three value propositions in mind, here is an all-encompassing one which aims to appeal to all three of the different customer personas:

"Everything-Gym has the best facilities to help you lose weight, build strength, or get back into shape. Our diverse team of coaches have over 50 years of experience, so whatever your goal, we have a specialist on hand who is right for you."

Even if you don't intend to write and use a unique value proposition for each customer persona, you should still write one for each one anyway to help you to clarify your ideas. This way, when you write your all-encompassing value proposition, you can consider all of the customer types and write something you are confident applies to all of them. And if your tech idea is an internal tool rather than a customer facing product, the value proposition is how your team or organisation benefits.

Writing value proposition messages aimed at different audiences focuses your attention when deciding what features and functions you should build into your product. And if you plan on targeting one customer type before expanding to another, then you may also want to develop features of your tech idea initially only for that persona, something we'll see in more detail in the next chapter.

However, you shouldn't just write any old value proposition message and be done with it. The way you write your value proposition may be critical to your success. I've seen apps and websites over the years that generate zero leads due to poor value proposition messages, then a small change in wording opens the floodgates.

Which sounds better to you?

Proposition A: "Everything-Gym has 12 treadmills and is open 24 hours."

Proposition B: "Everything-Gym has tons of equipment and is open for 24 hours a day, 7 days a week, meaning you can work out on your own busy schedule and get your workout done without having to wait."

Proposition A is much weaker than B because it lazily just lists features. But people don't buy features — they buy **benefits**. Sure, having 12 treadmills and being open all day is great, but why does it matter? Don't neglect the 'why'.

You can conduct a simple need-feature-benefit analysis to make sure you write a value proposition that focuses on the right sets of benefits.

Each of your features has a benefit you can tap into to sell your tech idea, and the features and benefits you choose will vary depending on the consumer persona you're targeting. By identifying your customer's needs, which features address those needs, and the benefits those features deliver, you can align your messaging's focus to your target audience.

Here is an example of the needs, features, and benefits for different personas of gym user:

Persona: Elite powerlifting athletes.
Need: Win a powerlifting competition.
Feature: Lots of weights and equipment available.
Benefit: Lifting weights makes you stronger and having plenty of equipment allows you to train with enough intensity, without distraction, to help you achieve your goals.

Persona: Working professionals looking to lose weight.
Need: To lose weight.
Feature: Access to gym facilities with treadmills and other cardio equipment.
Benefit: Doing regular exercise improves heart health and the number of calories you burn, helping you to lose weight faster, and feel healthier too. This lowers your risk of cancer and increases life expectancy.

Rather than write these down in a list, get a sheet of paper and split it into three columns. Label the columns "Need", "Feature", and "Benefit", like this:

Need	Feature	Benefit

Reflecting on your idea, think of as many items as possible to fill the need column. Consider all possible needs for every type of customer.

Once you've got an exhaustive list, complete the feature and benefit column for each need. If any need has multiple different benefits, then create a new row for each.

When you have a list of over 10 items, review your list and you'll notice that each benefit will align with a particular customer type. Let's add two more columns to better record this correlation: Persona and Priority.

Need	Feature	Benefit	Persona	Priority

Write down each customer persona that applies the benefits you've identified. Here is an example where the same need has identical features and completely different benefits, each of which appeal to a different persona.

Need	Feature	Benefit	Persona	Priority
Lose weight	Cardio equipment	Better overall health	- Office workers - New mums	
Lose weight	Cardio equipment	Make weight for a bodybuilding show	- Body builders	

Each persona should be the profile of a customer or user, though you should avoid creating personas based on their roles, such as job role or label, and instead try to categorise them by their demographic, psychological traits, or other characteristics such as behaviours, values, or goals. [112]

Our analysis enables us to see that overweight office workers, post-natal mums, and body builders all want to lose weight but for different reasons. Having a different reason usually means there is a different benefit to the type of customer too.

Once you've matched your needs, features, and benefits with your consumer personas, you can start to prioritise them. Each company will do this differently, and there's no right or wrong answer — you have to decide the best option for your business. Some companies may create an overall value proposition that ties all of the persona's needs together, while others begin by focusing on one persona for their main value proposition.

For example, you might start designing your value proposition by focusing on students and align your features towards this persona. You can then focus on achieving results in the student market first, making the most of your limited time and resources before you expand to target new personas once your initial successes gives you more time, money, or resources.

So, how do you decide which set of benefits and value proposition is right to focus your attention on first?

One way is to consider the largest, most impactful group and start there — going for the low-hanging fruit. You could start with the largest of the audiences, or the one likely to spend the most money.

For example, let's imagine that your tech idea could appeal to more than one persona at the same time. And let's assume that the two personas your features and benefits best align with are the student and mum personas. If there are significantly more students than mums who are interested in your industry, it may make sense to target students first and focus your primary value proposition on the benefits students will receive because they are the larger audience.

But is this definitely the best prioritisation strategy? What if you discovered that although there are fewer mums than students in your addressable market, mums spend more on average? It might make more sense to aim your value proposition at mums despite the larger student audience.

Example

If there are 20,000 students you could target, and the average student is willing to pay £10 for your offering, then your total addressable market for this persona is £200,000:

$$20{,}000^{\text{students}} \times £10^{\text{average spend}}$$
$$= £200{,}000^{\text{addressable market}}$$

If there are half the number of mums you could target, 10,000, then you may worry that the market is smaller. But the market is not smaller if they're happy to spend more.

Let's assume that mums have more disposable income than students and are therefore happy to spend £100 for a similar offering. Although the addressable market is half the size based on the number of people in it, it's five times the size in revenue opportunity:

$$10{,}000^{\text{mums}} \times £100^{\text{average spend}}$$
$$= £1{,}000{,}000^{\text{addressable market}}$$

The size of the addressable market isn't the only factor that can influence your choice, and you may still decide to go for the smaller market if you think you can compete more easily. For example, if you were a student, then you have a unique competitive advantage in targeting them first because you probably know more students and have more domain experience to target them.

There are other factors that influence which value proposition you target first, such as how often each persona buys, their loyalty to companies, their likelihood to recommend you, or how long they are likely to stay a customer.

You need to be able to make decisions on where to focus your attention and resources first, and that's why it's critical to do a need-feature-benefit analysis before writing your value propositions. You're likely to waste a great deal of time, effort, and money if you aim your tech idea in the wrong direction, meaning an hour's planning now could save hundreds later.

If you found the need-feature-benefit table useful, you can download a free template file from the Execute Your Tech Idea website to help you to define your value proposition.

Free Resource Pack

The Execute Your Tech Idea website contains further information, quick-start document templates, and other helpful free resources

executeyourtechidea.com

The customer isn't always right

A few years ago, I worked with a client in the events organisation space. We were helping them to develop software that already had several active users. As we continued to work on the features that best met the client's strategic priorities, they had a constant competing force distracting them — their clients requesting new elements be added to the interface.

It's incredibly hard for any business to ignore ad hoc requests like this, and customers might request these changes loudly! It's tempting to drop everything and implement the requests as soon as possible, and sometimes this is the correct thing to do, but not always. Anything you choose to do now is at the expense of another thing that you could have been doing now.

So, contrary to that popular cliché phrase, the customer isn't always right!

Like caring for the wounded in an accident, you may be tempted to go to the ones who scream the loudest, but it's the unconscious ones who aren't crying for help who often need your assistance first. Anyone you save now is at the expense of saving someone else now in their place.

Our client would have got much better results if they had outlined a clear value proposition *before* developing their new customer's requested features, and this strategy would have allowed them to find the ideas that best aligned to their priorities.

Armed with this value proposition, they would have been able to identify features that worked better for the client, understand the client's needs, and keep their active users happy whilst not straying from their vision. Fortunately, I was able to convince them to adopt this more considered and less reactionary approach.

Another historic example of a company that isn't afraid to put its values above the customers' requests is Apple.

Despite fierce criticism from their customers, Apple removed the headphone jack from their flagship smartphone. Few customers had asked Apple to ditch the jack, as it was a valued feature — who would?

But Apple knew they needed to focus on higher priority features and benefits that better aligned with the brand's values and its market, such as battery life and a sleek phone to drive the company forward, contrary to what their users thought they wanted. Apple let their vision guide their values, and their values drive their priorities.

This is a great illustration of the fact that your value proposition shouldn't only drive what you choose to do — it should also drive what you choose *not* to do. And, what you choose not to do is sometimes more powerful.

Closing words

Now you have a clear framework on how to identify the needs, features, and benefits of your tech idea, you can consider these alongside your vision to identify a value proposition for your business or the tech initiative you are looking to execute.

But be careful — if you have one value proposition for your overall business and more narrow value propositions for individual products, processes, or features, make sure they don't conflict with one and other. For example, if the main value proposition of your business is to deliver quality above all else, then it may not be sensible to introduce a product range with a value proposition of low cost at the expense of quality!

Key takeaways:

1. Make sure you consider the value proposition of everything you do, as without one, you are shooting in the dark.

2. Use a need-feature-benefit analysis to identify your value proposition or the value proposition for a particular persona.

3. If you have multiple personas with competing needs, focus your attention to create a value proposition that works for all of them. If you can't create an all-encompassing value message, then target the personas with the highest opportunity value.

4. Use your value proposition to guide your actions, decisions, and priorities.

5. Remember that your value proposition is equally important in deciding what *not* to do. Try to do the right things, not all things.

CHAPTER 7:
YOUR MINIMUM VIABLE PRODUCT (MVP)

"As you consider building your own minimum viable product, let this simple rule suffice: remove any feature, process, or effort that does not contribute directly to the learning you seek."
— Eric Ries, Author of "The Lean Startup"

Chapter Relevance	
New Internal Tool	**New Product**
↑	↑

In this chapter, we'll look at the Minimum Viable Product (MVP) concept, which is a philosophy for how you should decide what to build, when, and in what order.

We'll outline what an MVP is, how it relates to the vision and values discussed in the last two chapters, and how to prioritise what you do and when you do it. Every decision you make about priority is based on your assumptions and existing biases, so we'll end the chapter with ten methods to validate your MVP features.

What is an MVP?

The core idea behind planning your MVP is aiming to build and launch the *simplest* version of your tech idea as possible. The goal is to reduce waste, speed up your launch, and test your assumptions before building on them further.

It's tempting to aim for perfection and release your tech idea with an extensive, all-encompassing list of features. The logic behind this is: the longer you spend on development, the more value your idea brings and the more likely it will be successful.

This logic might be true in a world with no resource constraints, unlimited time, and an endless budget... but in the real world, it's highly dangerous. What if aiming for perfection means taking six months longer and costing twice as much, only to realise that half the features you thought customers would value aren't even used. Perfection doesn't sound so perfect anymore, right?

Instead, the MVP philosophy requires you to be laser-focused on what you build and in which order.

By this point, you probably have a rough idea about the different features and functions that could go into your tech project. Each of these features take design time and effort, development time to implement, and project management time to coordinate and test. Any feature you build that you later find you don't need is wasted time and money. So, ruthlessly cut the non-essentials to spend your limited time and money as efficiently and as effectively as possible.

However, building an MVP does not mean outputting lots of rubbish quickly, and you aren't aiming for the smallest project, but the smallest project 'possible'. So think carefully how you balance your resource constraints with the values of your project. For example, you may make your first project version slightly larger than you otherwise could to deliver the right amount of initial value when you release it to keep users interested.

The goal is to recognise that each new feature costs time and money and delays your launch date, so optimise your plan of execution around this. Decide how big or small your end product should be based on your priorities and available resources.

For example, imagine your perfect project consists of 12 large features, each of which takes exactly a month to build. Which sounds better?

1. Take 12 months, build all the features, and then launch in one year's time?

   Or

2. Build six of the most important 12 features first, launch in just 6 months, then spend the next six months building the remaining features whilst version one of your MVP is used by real people, giving you valuable feedback from early adopters?

In most scenarios, the second option is preferred as it gives you the best of both worlds: you get to launch sooner, for less money, and still work towards building the perfect product in the original 12-month timeframe.

When explained this way, the MVP philosophy seems like a no-brainer, but in practice, it's not always a clear-cut decision. For example, you may think: why not launch something even sooner, in just a month? You might be able to, but does this approach allow you to meaningfully test your initial assumptions, and is the end product functional or fit for purpose?

Your unique perspective

As we discussed in earlier chapters, your innovation ideas will come from your sight or experience of a problem that exists within the market. This unique perspective on the world allows you to connect the dots in ways that seem obvious to you but might not be apparent to others.

Take my wife Aimee for example. [113] She's a Unit Stills and Specials Photographer who takes photos on set of big-budget film and TV productions. [114]

In her day-to-day job, Aimee could see issues with the way that actors approved the images she took. It turned out there were so many stakeholders involved in the image approval process that it became impossible to manage. There were actors, agents, PR departments, publicists, producers, studios, productions, directors, and photographers, as well as common contract terms and image secrecy issues to navigate.

Some film productions tried to solve the problem with complex file and folder structures on their computer, dragging and renaming files. But with tens of thousands of photos for most film and TV shows, these productions quickly found themselves drowning in admin and coordination work.

Following the MVP principles, we started planning and developing a platform that simplified the whole process. Our first features included users accounts for actors and their agents, a back-end area to add productions and upload images, a process to allow actors to "kill" images, and simple reports that allowed production to see which image assets were safe to use.

Given Aimee's job, she is in the privileged position of being both the creator and the end user of the tool we set out to build, ImageApprovals. [115] Her knowledge of the market allowed her to decide which features were most important to build in which order. However, there were still some grey areas where it wasn't easy to decide how or what should form the MVP. One of these areas was the billing function.

If you're selling a service online or within an app, you need a way to collect payment. Making money is very important to the success of a new service launch, so collecting payment automatically becomes a priority feature that should go into most MVP launches. Surely with ImageApprovals being a software service, online payment capabilities were critical, right?

Not in this case. Or at least we decided not to implement automatic billing functions for phase 1. Why? Because there were lots of stakeholders involved in the image approvals process, and each one of those stakeholders had the potential to be the buyer.

Given the complexity of different billing contacts in the payment and sign-up process, we decided to invoice each account manually for the foreseeable future. In practice, we found that customers buy the service one or twice per year, which meant the decision didn't create too much admin work, and had the benefits of allowing Aimee to control the sales process herself.

Another benefit of conducting the process ourselves was that we spotted unforeseen improvements. Once we'd gone through the manual contract creation and invoicing process many times, we noticed similarities in how each stakeholder wanted to use the system. By the time the manual process became repetitive, we'd found the right way to do it and were in a much better position to build an automated payment function into the software.

So, although you might have to cut something out of your MVP for now, it doesn't mean it's gone forever — you can always add it back in when the time is right.

Prioritise your MVP

Now you know what an MVP is, you might be wondering how to decide which features to include and which to exclude. You certainly shouldn't cancel or approve features based on your gut feeling alone. In this section, we'll look at how to prioritise features using a technique called the impact-effort matrix.

Think of impact as the level of business results, and effort at the total difficulty or cost.

Here are some criteria to think about when deciding which features should form your MVP. Each of these questions will help you identify the effort to implement a feature and the impact the feature has on your business:

1. Will people use the feature regularly? (Impact)
2. How long will it take to develop? (Effort)
3. What is the cost to build it? (Effort)
4. Will it need to be removed or replaced later? (Impact, Effort)
5. Does it add enough value? (Impact)
6. Why should the customer value it? (Impact)
7. Does it support our business goals? (Impact)
8. How does it support our business goals? (Impact)
9. Will people use it immediately or in the future? (Impact)
10. Is there a manual approach that can be replaced later? (Effort)

11. Is this the right time to launch this feature? (Impact, Effort)

12. Does is complement our brand values? (Impact)

13. Is there a simpler way to solve this problem? (Effort)

14. Is there a risk to the business of implementing this or implementing it this way? (Impact, Effort)

If a requirement or feature takes longer, costs more, or is risky or difficult to implement, then the effort is higher than those that take less time, cost less, or are easier to implement.

Impact is a measure of how important the feature is towards meeting your business goals or your clients' goals. If a feature or requirement helps you achieve a high-value goal, more revenue, better customer experience, or strong business results, then its impact is higher than features that deliver fewer of these values.

Here is an example of an impact-effort matrix. Each number represents a feature or strategy that could form part of your MVP. The position of each feature on the matrix is based on its impact and effort.

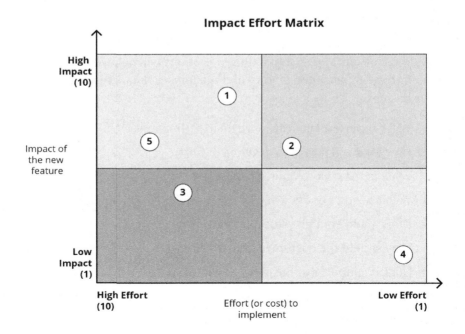

In this example, you can see that item #2 is low effort and high impact, and #3 is high effort and low impact, therefore #2 should probably be prioritised.

Previously, we discussed the importance of conducting a need-feature-benefit analysis to identify the value proposition of your idea for your target customer (or for internal tools, your target user or business). The process of arriving at a value proposition is important because it will guide most business decisions you make, including how to score items on the impact-effort matrix.

For example, imagine we have an existing business in the fitness industry, ABC fitness. The business provides coaching to athletes to reach their strength goals, such as lifting heavy weights or performing well in a competition. The owner, Big Danni, conducts a need-feature-benefit analysis and comes up with the following value proposition:

"ABC fitness helps aspiring athletes to meet their strength goals and out-perform their previous competition results. By pairing top-performing coaches who also compete with aspiring athletes, combined with 10 years of scientific research, we have a track-record of setting training programs that boost the amount you can lift by 5–10% per year, on average.

ABC fitness is generating great results for its clients, however as the business grows larger, it finds managing hundreds of clients each month manually is inefficient and causes administrative problems. These problems hurt the customer experience, making it hard for Big Danni to scale his business. If Danni doesn't first fix these problems, then they will multiply as the business grows.

Big Danni sees an opportunity to streamline the business with a mobile app that will also act as a unique selling point for ABC Fitness, enabling them to win more customers. The goals of the new app:

1. Reduce admin time spent each week managing the process.
2. Improve the overall experience for athletes going through the ABC fitness process.

Excited about the prospect of launching a new app, Big Danni writes down his ideas for the features that it could contain: [116]

1. User accounts for both athletes and coaches.
2. In-app messaging system.
3. A process to allow athletes to record their gym activity.
4. A process to allow coaches to set a weight-training program for the athlete to follow.
5. A 'share results' function on social media.
6. In-app payments to collect monthly subscription fees.
7. Performance graphs and metrics for the athlete.
8. Push notification news updates.

Now let's evaluate how each of these features relates to the value proposition of ABC Fitness and Big Danni's business goals, scoring each item on the impact-effort matrix.

1) User accounts for both athletes and coaches

The commercial value of the app is to enable users to log in and interact, so there's no point implementing the app without this function. The size of this requirement is also not that large, taking 1-2 days to implement.

Impact: 10/10
Effort: 3/10

2) In-app messaging system

This function allows the athlete and the coach to message each other. This sounds like a simple requirement, but the development team tells Danni it will take a full week to implement, plus testing. Though this is a nice feature to have, the purpose of the app could still be fulfilled if the coach and the athlete used another messaging app in parallel, such as WhatsApp or SMS messages.

Impact: 3/10
Effort: 7/10

3) Process to allow athletes to record their gym activity

Danni wants to sell the app to coaches so they can easily set programs and track their progress. The recordings of the completed programs for the athlete are very useful and add a lot of value for both the coach and the athlete. The app could technically launch without it, but it might be harder for Danni to sell the benefits without it. The development team tells Danni the time to implement this feature is three weeks plus testing.

Impact: 9
Effort: 9

4) Process to allow coaches to set a weight-training program for the athlete to follow

As the main goal of the app is to allow coaches to set a training program for the athletes they coach, without this feature, the app does not add its intended value. The development team tells Danni this feature will take 2 weeks to implement.

Impact: 10
Effort: 8

5) Share results function to social media

Sharing functionality may seem like an easy requirement that can be cut out without impacting on the value delivered by the app, but Danni hopes that the coaches and their athletes will like the app so much that they share their progress with others. His marketing strategy hinges on the plan that sharing will help the app to 'go viral' and allow the business to do more with a lower marketing spend.[117] The development team says this feature will take two hours to implement.

Impact: 9
Effort: 1

6) In-app payments to collect monthly subscription fees

Automating the online payment process within the app would save a lot of time, but Big Danni thinks he could begin by invoicing customers directly. With this approach, he would set up a direct-debit mandate himself to take recurring monthly payments. This has a trade-off, however, because as the project scales, this manual process will be difficult to track. The development team quotes 3 days to build this feature.

Impact: 2 *(Though the later business impact will be greater)*
Effort: 5

7) Performance graphs and metrics for the athlete

Having insights into how the athlete's performance has improved over time allows the coach to update the program each week. Without this data, it's very difficult for the coach to be effective, reducing the value of their service to the end customer, the athlete. After discussing this feature with the development team, they suggested that there was more than one approach to implement this feature.

Approach A

They could implement all six graph types, from pie charts to histograms and line charts, giving both athlete and coach the ability to specify a date range for the data returned on each graph. This is the full bells-and-whistles approach and would take two full weeks to implement.

Impact: 8
Effort: 8

Approach B

They could implement one of two simple line graphs to show the data and fix these graphs and only show the last 12 weeks' worth of data. There would be no ability for the athlete or coach to select a date range with this option. This balanced approach gives users the most important data and reduces the amount of development time required compared to approach A.

Impact: 7
Effort: 4

8) Push-notification news updates

Though notifications may be a good way to remind the user that the app exists (thereby encouraging more use) sending news updates is not a core part of the value proposition of the app. Not all users will engage with the feature. The development team says this will take two days to implement.

Impact: 1
Effort: 4

***

Now that we know the impact and effort to implement each requirement, we can map this on our diagram.

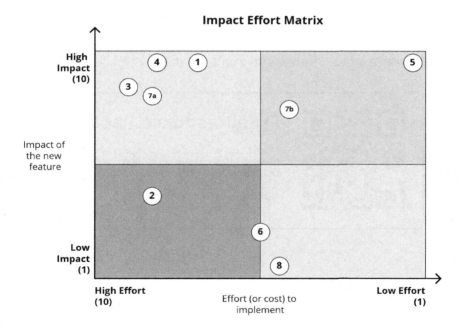

This impact-effort matrix allows us to see which features are worth implementing and which you may consider scrapping or developing as part of a future release.

My interpretation of this matrix is that Danni should probably prioritise feature #5 (social sharing) and #7b (simple version of graphs) over features #2 (in-app messaging), 6 (monthly subscriptions), and 7a (the complex version of graphs). This is because their impact is high and their cost is low.

Your requirements and impact-effort matrix will change over time. Things you thought would be a priority one day might not be the next, especially once your product is in the hands of real users.

This ongoing shift in your perception of priorities is why I recommend that you keep your MVP as small as it can be to test your assumptions rather than trying to scope, design, and plan a 12-month project entirely upfront. Chances are that six months in, you'll want to change your plans, which I'll elaborate on later and give you project management techniques in "Chapter 13: Project Lifecycle Types" and "Chapter 17: Plan Your Product Roadmap".

If you found the impact-effort matrix useful, I've created a free template file on the Execute Your Tech Idea website, which you can download and use to map your own features onto the matrix.

Free Resource Pack

The Execute Your Tech Idea website contains further information, quick-start document templates, and other helpful free resources

executeyourtechidea.com

Validate your MVP

Validating your MVP is the process of testing whether the features you have planned are the right ones. But why is this important?

The planning, managing, and programming of your tech idea after you've defined your MVP is time-consuming and costly. And turning your ideas into reality will take hundreds of hours and several months of project and development time. So, before you leap in and implement the features you've now prioritised on your impact-effort matrix, it makes sense to test whether your priorities are correct! You must validate your assumptions.

If you think a feature is worth paying to build, ask yourself why? You've categorised it as high impact because you think user will love it. But if you've not asked them directly, how can you know for sure? What if you show them the feature and they hate it? In this scenario, the impact score should have been set much lower than you initially believed as your starting assumptions were incorrect.

To validate your MVP priorities, you must expose the assumptions you've made to the world to see whether they are true.

This process of validating your MVP makes implementing your tech idea much less risky. A small amount of initial real-world research will prevent you from wasting time and money implementing things that people don't want or implementing things in a sub-optimal way.

With that lesson in mind, here are ten methods you can use to validate your ideas and check whether your MVP plans are realistic.

You should spot similarities and crossovers in these methods with some of the idea generation approaches mentioned earlier in the book. This is not a coincidence, and it's common for techniques at one point in your journey to add value to other points.

Method 1: Landing pages

Landing pages are a powerful way to test that your value proposition will land well with your target audience.

Set up a landing page on a popular website building platform, present your value proposition, and upload images or prototype pictures. [118] Allow people to sign up to try the early-adopter version for free when it's available.

Now you can promote your idea using the website and see whether anyone signs up. If you don't get results, you can experiment by changing your marketing message and value proposition until you find an approach that works. It's vastly cheaper to change your ideas and messaging now than change features in your product after it's built.

Method 2: Customer surveys

Customer surveys can be a brilliant source of information to test your MVP assumptions and allow you to uncover things that you didn't know that you didn't know.

You can create a survey with specific questions that relate to your product, starting with the most basic: do you like it? To collect honest feedback, ask people in your target audience to fill out the survey after seeing your prototypes. Gather your surveys together and look at which points keep popping up before using them to develop your product in its next stage.

If you haven't made a prototype of your tech idea yet, ask your audience questions that will help you develop it. For example, if you were looking to create a mobile banking app, you could ask what issues they have with their current bank, what they love about mobile banking, and which services they use most frequently.

Customer surveys are used by businesses both large and small around the world. Many large companies, for example, use NPS (Net Promoter Score) surveys to generate feedback on their products. [119] They'll ask questions like 'How likely are you to recommend the service to a friend or colleague?' and users will give them an answer to choose from between 1-10. They use this information to judge how happy consumers are with their business and get valuable insights into the success of their products.

There are plenty of tools out there to help you create professional surveys. [120] These tools allow you to create your own online form containing different stages and questions that you want to ask. Once a user submits the form, the data is collected on a spreadsheet or internal database. You can then reflect on this feedback to see whether there are any common suggestions for improvement or areas where people encounter the same issues or benefits.

But be warned! You should think very carefully about the questions that you ask in your survey. If you're ambiguous in the way you word each input, you may confuse users and not receive useful responses. You should also be careful not to ask leading questions that might bias the result, for example "How would you score your experience?" might be a better question compared to "How amazing were we?"

Method 3: Wireframes and mock-ups

Wireframes and graphic-designed mock-ups are a form of proof of concept. These prototypes don't have to be real, usable functions of your product, but simply something that demonstrates that your idea works, has an audience, and can be developed further. With these prototypes, your aim is to model your tech idea without having to program the whole thing.

Here is a quick example of the difference between a wireframe (left) and a user interface mock-up (right):

Wireframes and graphic designs are another common form of prototype if your tech idea requires an app or online portal. These mock-ups demonstrate what each screen could look like, and you can create clickable prototypes that give you an idea of what the app might feel like for the end user.

Use wireframes in combination with the other validation methods for maximum impact.

Method 4: Do it manually

You can also create a manual prototype and do everything yourself in person. This is where you manually handle the business processes that you eventually want your final implemented tech idea to replace.

To take advantage of this manual prototype approach, think about the most important functions of your tech idea, what value they bring, and how the user interacts. Be creative to see whether you can identify ways to do these tasks manually, such as calling users to collect information or sending updates via emails that you write yourself.

Though time-consuming, taking a manual approach to prototyping your processes means you can experiment to see whether the basic process works without technology first. Manual prototypes also help you discover customer requirements that you didn't expect or edge-cases in the process that you'd not catered for. This is because while conducting the phone calls yourself, you will see first-hand where the customer gets confused.

Once you've manually tested a process for a while, you can then adjust your approach based on what you've learned. This way, once you automate things with tech, you know you're coding up processes that already work.

Method 5: Forms and spreadsheets

Forms and spreadsheets are to tech ideas what duck tape is to DIY. They help you bodge together a prototype process, which you can replace with something more sophisticated later. The beauty of this approach is that if your ideas don't work out, they are cheap to scrap and start again. Think of this method as one step up from manual prototypes.

For tech ideas that require data collection from users, you can test your ideas through a combination of cloud office tools such as online forms. [121] You can use these tools to collect the information that your tech idea will also need to collect and store in a spreadsheet or online database.

A spreadsheet prototype is slightly different to just storing information from forms — it's where you use formulas, function, or scripts in your sheet to demonstrate elements of your tech idea. For example, if your tech idea has a complex calculation, you can create a demo version of that calculation in a spreadsheet first.

I once had a client who developed a sports training app that aimed to help athletes jump higher — a useful skill for basketball, volleyball, and many other sports. [122] When the client first asked me about app development, they showed me a spreadsheet they had created and refined over the years. It contained lots of inputs for training sessions, exercises, reps, and weights lifted, and they gave it to their athletes to complete every week. The spreadsheet used formulas throughout to automatically calculate next week's training plan from last week's performance.

Our jump coach client had been working with a prototype of the app for years without even knowing it! Their prototype showed us a few things — they had an audience, the concept worked, and they knew which calculations the app should replicate.

If your business processes require lots of steps or complex calculations, maybe you can create a prototype spreadsheet for your tech idea too. This will help you to map out your ideas and will better enable you to show what you aim to achieve to others.

Method 6: Cross-reference search trends

Did you know that at the time of writing, the Google search engine is used over 3.5 billion times per day for different queries? Wouldn't it be great if we could see what people are searching for and use that to influence our product, services, and marketing?

This is exactly what search-trend tool Google Trends enables you to do. [123] It makes analysing this real market data easy. Simply type in any search term that's relevant to your tech idea and it'll bring up a wealth of information about that term, including interest over time, interest by region, and related searches.

By looking at what people are searching for, you can get a better idea of what's relevant in your industry and which features might be a hit amongst your audience. Trend data might also help you optimise your efforts to focus on rising trends rather than declining ones. As such, it can be a useful tool to help you focus your priorities and targeting. A declining trend may make you question whether there is demand for your idea, and a growing trend might be an indicator of an unmet need that your idea solves.

Think about what language you would use in your product descriptions and landing pages, then use Trends to analyse those keywords. You may find there is little search volume for some keywords, which may indicate a small market size. Or you might see substantial growth in certain keywords or industries, giving you confidence that your tech idea can take advantage of the rising opportunities.

Method 7: Analyse competitors

While you don't want to copy and paste a competitor's business, that doesn't mean you can't use what they're doing for research purposes.

When validating your MVP idea, investigate what other businesses in your industry are doing. Try online searches of competitor companies to analyse their marketing, branding, and products. Get extra information by signing up to their mailing list for new product launches and details.

Think of most things that you buy or use, whether it's your phone, the software on your computer, or even the brand of toothpaste you use. Chances are there are tens if not hundreds of alternatives that you could choose instead, but you use the products that you've chosen — why? Because of the unique value they bring relative to the other options.

So, if a company is offering a similar product to your idea, this isn't necessarily a reason to be put off. Instead, use it as proof of concept — evidence that your idea could work in the real world and that a market exists. This adds value to your own product and could even increase your chances of buy-in from your company or investors. If a market exists and you think you can own a unique niche, then you're in business!

Another check to carry out when researching competitors is who's getting funded and by whom. Look at investing directory websites, which show new products coming to market and who is investing in them. [124] Read start-up funding blogs to see which companies are drawing the most attention and why. If you see companies that are similar to yours getting funded, it's a good sign that demand exists, or at least that others believe in the opportunities in the sector.

Of course, look at your competitors' websites and get an idea for what they do. If they have a narrow product or service offering, this is particularly useful as their business performance is more likely to be related to the strength (or weakness) of this narrow offering.

Another way to research the competition is to do some due diligence on their accounts and financial information. You can get some company information by looking them up on the government database, which in the UK is Companies House. There are also paid-for tools that allow you to perform more detailed reports. [125]

Method 8: User demos

You don't need a fully-fledged working product to get your tech idea out there. User demos build on the wireframing and mock-ups method we discussed earlier but involve manually demoing what you've put together to a prospective user and talking them through the process as if it already exists.

Start marketing your demo graphics across your marketing channels, including on social media and your website, and offer a compelling discount to people who agree to sign up before the product even exists. Then if people show interest, follow up and invite them to have a product demo where you talk them through your ideas and prototype.

You can use an email marketing tool to populate a list of early adopters so you can email them later when your platform is complete. [126] You can also encourage early sign-ups via a landing page with an embedded form to see what reaction you get. Consider running a poll to generate a buzz, user engagement, and find out people's opinions about different elements of your idea before you build them.

For in-person demos, you should combine direct and indirect forms of collecting feedback. Start with an indirect approach by showing users what you've created with little to no other prompts, allowing you to gather their initial thoughts without you biasing their opinions and responses.

Next you can get slightly more direct, asking them leading questions to get answers. You might say things like "what would you expect of this?", "what does this wording mean to you?", or "what is your opinion of this section?".

Finally, once you've collected their unbiased feedback, you can ask more direct questions like "do you think the target audience would buy this?" or "do you find this button clear or confusing?" You save the direct questions until the end because if you ask them at the beginning of the demo, it can bias answers through the rest of the process. Align the questions you ask to the assumptions that underpin your need-feature-benefit analysis, value proposition, and impact-effort matrix.

Method 9: Internal feedback

If you're building an application for internal use in your organisation, then you're lucky enough to have your target audience right where you want them. All you have to do is share your ideas and ask your staff what they think or apply earlier lessons in this chapter by creating a survey for more insightful feedback. After all, they will ultimately be the users of the tool once it's built.

Internal users can help you learn more about their needs and whether they think the product will be practical for its use. However, be careful not to succumb to every demand from your audience and instead look at the consensus is across the board or whether the feedback aligns (or conflicts) with your priorities and value proposition.

You can then combine the techniques used in the other methods to gain feedback from these internal stakeholders.

For example, if you've created a spreadsheet or prototype, you can arrange an in-person demo to see what potential users think and whether they have any suggestions to improve your ideas. You may also find opportunities to collaborate with other departments, extending your tech idea to provide more value by collaborating than you thought possible. Or you might simply ask a leading question of a colleague in the coffee room. The validation methods don't need to be time-consuming or costly as long as you get the information and learning that you desire.

Method 10: Call prospective customers

The most powerful way to validate your idea is if real customers buy it with real money.

Some businesses with established tech products generate most of their revenue by having salespeople, sometimes the directors of the business, call prospective clients. They will call, generate interest, and if possible, get the customer to buy the product or service on the call. [127]

If these salespeople can conduct the whole call and get buy-in with nothing but a phone conversation, then why do you need the product to exist to start selling it? You don't need a real product to test whether your sales process will work.

Cold-calling has got a bad reputation these days, but there are ways you can do it without tarnishing your brand image. [128] Just as you have people sign up for mailing lists, you can have your audience add their phone number when buying other products from you or when showing interest in your company via your website.

Plan a sales script and start calling these people and pitch them your MVP as though it already exists, gauging the reaction of the potential client. This is a little different to a user demo, as you're pitching the description of your features, benefits, and value proposition rather than showing them. This can make this technique useful for very early-stage validation of your ideas where you haven't yet created anything.

Just be careful to see the difference between lukewarm interest and someone who wants to buy. For example, someone who wants to buy is likely to ask more questions and be more precise with positive feedback, whereas lukewarm interest is vague and non-committal. To test your assumptions, act as though you're trying to close the deal. If the caller is interested, let them know how they can stay updated about your product as it becomes a working MVP.

You'll gather valuable feedback about all elements of your target business model and tech idea. You'll spot areas to change and new opportunities that should take your focus and become a priority. Make sure you write down any common objections you discover to adapt your ideas.

Closing words

Sometimes an MVP requires you to go all-in and build a functional version of the product for real. Sometimes it means a more basic prototype to help you test your ideas. The approach you take and amount of effort to build your MVP will vary depending on the constraints and pressures you experience in your business, and you will need to use your best judgement.

For example, if you have financial or deadline constraints, you may have no choice but to plan and build a much smaller MVP than you would otherwise.

You may also be forced to build a large MVP if what you do already commands a large 'minimum' set of requirements. For example, if your tech idea is a replacement for another system or process, it may not be possible to retire the old system until the new one performs all of the basic functions of the old one, which might require you to build a much larger MVP than for a completely new process or product.

Remember, creating an MVP isn't about trying to make your first phase of development as small as possible — it's about making it as small as you feel it *reasonably* needs to be to meet your objectives. This sometimes means you can't avoid your MVP being big, but if you're planning a large MVP, make sure it's large for a good reason.

Once you're clear on the objectives of your business, consider your vision, value proposition, and budget to start deciding which features should be included in your MVP and which should be discarded or delayed for future phases of the project.

You should then identify the main features of your project and map them on the impact-effort matrix template provided earlier to identify the order of priorities based on the value that a feature brings relative to its effort or cost. You may choose to do low-effort and high-impact activities first.

Once you've defined your value proposition and impact-effort matrix, you're ready to begin validating your assumptions using the ten methods. You can then re-visit your value proposition and matrix based on what you learned from your validation activities. You might choose to de-prioritise or remove some features or implement other new ones. Either way, you can be sure that your ideas will improve each time you test them.

Key takeaways:

1. A Minimum Viable Product, MVP for short, is the philosophy that you should build the leanest version of your ideas as is reasonably possible first to test your most important assumptions before building on them.

2. You can use an impact-effort matrix to decide which features should make up your MVP or which order to implement your ideas in.

3. Your MVP plan should consider your vision and value proposition to guide your priorities of what to do and when.

4. Make sure you validate that your assumptions around priorities are correct using the ten methods.

5. Be careful not to bias the feedback you receive when validating your ideas. Ask open questions first before moving to more specific ones.

6. Re-visit your value proposition and impact-effort matrix to make adjustments after attempting to validate your ideas.

CHAPTER 8:
RAISING AND MANAGING MONEY

"If I had to run a company on three measures, those measures would be customer satisfaction, employee satisfaction, and cash flow."
— *Jack Welch, Former General Electric CEO*

Chapter Relevance	
New Internal Tool	**New Product**
→	↑

Raising and managing money is critical. It's one thing to have a high-value idea that is obviously worth the time and money investment to implement on paper, but you are stuck if you don't have the resources or funds to complete your project.

In the last chapters, I provided the tools so you could decide how to add value and use that value proposition as a lens for what you should build and when. However, you can't get away from the fact that the amount of money you have available will heavily dictate your plans. If you don't have enough money, then you might be forced to deliver a more watered-down version of your idea than you would otherwise.

This chapter addresses the topic of money and provides methods to fund your project and make sure that funding is enough. These tools can help you to see when your project will become self-sufficient and pay for each month's costs directly from profit, a concept I call your 'time-to-money' position. [129]

Cash flow basics

Creating a cash flow forecast is one of the most important things you can learn in business, as it allows you to see when money comes in and out so you know whether you have enough money to execute your plan or need to raise additional funding.

A cash flow forecast is a running total of what your bank balance looks like each month, alongside sources of money-in (income) and money-out (expenses).

As we're in the planning stages of your tech idea, I'd like you to make your best guess of what your income or expenses might be based on your gut feeling. You will be able to swap out these guesses with better estimates as you progress with your idea.

First, plot your months, quarters, or years in a sequence like this:

Example

Month:	1	2	3	..etc

Each column represents a calendar month, which I have labelled as 1, 2, 3, and so on. You could easily label them Jan, Feb, March, depending on what month it currently is.

Next, record your income against each month.

Example

Think about where your income will come from and have a new line for each source, then predict how much money you will get each month against that source.

For example, if you expect to have £1,000 in income every month from sales, plus £100 from other places, then put this into your cash flow table as follows:

Month:	1	2	3	..etc
Income: Sales	£1,000	£1,000	£1,000	
Income: Other	£100	£200	£100	
Month Income:	£1,100	£1,200	£1,100	

Notice how the bottom "month income" row ads up all of the incomes for each month? You can see that in month 1, you make £1,100 and in month 2, £1,200 because month 2 makes £1,000 from sales and £200 from other sources.

Once you've recorded all the sources of income, you can do the same for all expenses.

Example

At the top of the image, you will see the income items we added a minute ago, but if you look underneath, you can see I've added some expenses.

I've assumed that the development costs will equal £4,000 a month until development is complete in month 3. Then marketing can begin at the same time, costing £500 per month from this month 3 onwards.

I've also assumed there will be £100 each month in other expenses.

Month:	1	2	3	..etc
Income: Sales	£1,000	£1,000	£1,000	
Income: Other	£100	£200	£100	
Month Income:	£1,100	£1,200	£1,100	
Expenses: Development	£4,000	£4,000	£4,000	
Marketing	£0	£0	£500	
Other	£100	£100	£100	
Month Expenses:	£4,100	£4,100	£4,600	

Notice how the monthly expenses row at the bottom is a total of just the expenses for each month? For example, in month 3, there are £4,600 in expenses coming from £4,000 of development, £500 in marketing, and £100 in other costs.

There are now lots of rows and columns in this spreadsheet but it's actually quite simple stuff — we're just adding up.

Once you have a row containing the total for each month's income and a row with the total for each month's expenses, you can see whether you are making or losing money each month. Let's call this the month flow.

Example

In this next example, I've hidden the list of incomes and expenses and included only the total income and total expenses for each month.

You can see that in month 2, the income is £1,200 and the expenses are £4,100. This means the total month flow is the income minus the expenses, which is minus £2,900.

Month Income:	£1,100	£1,200	£1,100	
...	...	...	...	
Month Expenses:	£4,100	£4,100	£4,600	
Month flow	-£3,000	-£2,900	-£3,500	

In other words, you spent more money in month 2 than you earned, meaning if you had a bank account, it would have £2,900 less in it at the end of month 2 than at the end of month 1.

Finally, we can calculate the cash flow, which is a running total where each month's flow is added to the previous month's flow. It helps to imagine this as your running bank balance over time.

Example

For example, our cash flow at month 1 is easy to calculate — we can see that our income minus our expenses leaves us with minus £3,000. This means we have £3,000 being spent from the bank account, which we do not have.

Then by the end of month 2, you have spent a further £2,900. This means your cash flow at the end of month two is minus £3,000 from month 1 plus minus £2,900 from month 1, leaving minus £5,900. The negative bank balance has increased as you spent more money than you had for two months in a row.

Finally, we add another row to record this running total of your monthly bank balance, called your cash flow.

Month flow	-£3,000	-£2,900	-£3,500	
Cash flow	-£3,000	-£5,900	-£9,400	

In our example you lose money each month and you can see by month 3, your cash flow is minus £9,400, meaning your bank balance would be in debt by this amount.

Month:	1	2	3	..etc
Income: Sales	£1,000	£1,000	£1,000	
Income: Other	£100	£200	£100	
Month Income:	£1,100	£1,200	£1,100	
Expenses: Development	£4,000	£4,000	£4,000	
Marketing	£0	£0	£500	
Other	£100	£100	£100	
Month Expenses:	£4,100	£4,100	£4,600	
Month flow	-£3,000	-£2,900	-£3,500	
Cash flow	-£3,000	-£5,900	-£9,400	

Again, this looks like an overwhelming amount of information, but we're only calculating the total for each month, then the running total as the months progress.

A positive cash flow means that as each month goes by, your bank balance increases. A negative cash flow position means that your bank balance decreases. To stay in business, we must play an infinite game to prevent this number from ever hitting zero.

You can avoid hitting zero and losing the game by making more money in the form of revenue, reducing your expenses, or accessing funds via some other means, which we will cover shortly in this chapter.

Typically, most start-ups or new projects have a period where they lose money each month before they start to make money. They invest money upfront to build their technology and aim to make more money from that investment later.

Example

Let's look at the cash flow for a business with lots of expenses and no revenue in the first three months but starts making money by month 4. To simplify things I've removed all of the income and expense rows that lead to each month's cash flow figure.

						Positive cash flow	
Month:	1	2	3	4	5	6	7
Cash flow	-£5,000	-£7,000	-£7,100	-£4,000	-£1,000	**£2,000**	**£4,000**

Notice that the negative cash flow each month makes your bank balance go further and further negative until it peaks at minus £7,100 in month 3?

The cash flow then reverses direction, it becomes cash flow positive, and your bank balance improves each month until month 6 where it is above zero.

In practice for your tech idea, your outgoings will include costs such as development, marketing, advertising, graphic design, software costs, employee costs, and other expenses. Your incoming money may be from starting cash, loans, investments, and revenue. And for internal tools, this cash flow position will apply to your department's budget.

We'll cover the different ways to build your idea in "Chapter 11: Ways to Build Your Tech Idea" where you'll get a better idea about what your tech idea will cost to create. The method you choose will have an impact on your expenses and cash flow.

If you'd like a handy template to put together your own cash flow forecast, then you can download one for free from the Execute Your Tech Idea website.

Free Resource Pack ⬇

The Execute Your Tech Idea website contains further information, quick-start document templates, and other helpful free resources

executeyourtechidea.com

Time-to-money

The time-to-money position of your business is how long it takes to reach a point of cash flow self-sufficiency, meaning what it takes to achieve a positive bank balance after first becoming cash-flow positive where your income is regularly more than your expenses.

Example

Let's say you're employed and go into your first month of employment with £0 in your bank, get paid £3,000 at the end of the month, and pay £2,000 in expenses during the month.

You are surely cash flow positive, right? As you are paid £3,000, spend £2,000, leaving £1000.

Although overall, the month is cash-flow positive, as you spend all of the money before you are paid you are actually in a negative cash flow position for most of this month. On the penultimate day of the month, you've spent £2,000 that you don't have.

It doesn't matter that you'll be paid £3,000 shortly — if you can't raise £2,000 to cover the negative cash flow gap, then you're in trouble!

When you launch a new idea, unless you have an existing well-funded budget, you may need a plan to cope with periods of negative cash flow, where your bank balance is lower than zero.

You can have the best, most lucrative tech idea in the world, but if you hit zero and stay there, you'll go bust. So, not properly planning your time-to-money strategy might kill a fantastic idea before it gets the chance to get off the ground. But don't worry — I'll give you the tools that you can use to increase your chances of staying alive and solvent!

Here's an example of a business venture that has great prospects but a poor time-to-money situation. If not handled carefully, the time-to-money profile could risk its success.

The venture is a model-booking platform (human models, not plastic figures). On the surface, the business idea was simple: professional models sign up to the app to find modelling jobs and booking clients have access to an app to post their job, pay, pick their model, and arrange their shoot. The platform then makes money by charging a small fee on the booking, called the take rate. This model booking platform is a form of marketplace business.

The challenge that most marketplace businesses face is this: without both sides of the market — the supplier and the customer — you don't have a business. This creates a chicken-and-egg scenario.[130] How do you convince models to sign up before having any jobs on the platform, and how can you secure booking clients if there are no models available to book?

This is a tough problem to solve that takes time and creative thinking. You need to create plenty of value on day one to encourage your early adopters from both audiences. For example, this venture had a long period where the app launched with no booking clients, so the business encouraged models to set up an account and get their profile ready, earning points that would be relevant once booking clients signed up. After months of signing up models, the platform finally had enough users to launch the booking tools.

Although this strategy solved the chicken-and-egg problem, it didn't help the time-to-money situation. The whole time the platform was onboarding models, it has costs, such as development costs, without making revenue in return.

Even when the product launched, the fees made for each sale were only a small portion of the overall sale, which made it difficult to earn enough money to re-invest existing revenues in marketing to grow the business organically.

Example

Let's say that it's £600 to book a model. If the booking fee is 10%, you stand to make £60 for each booking.

But if it costs £70 in marketing to acquire[131] the customer, then the business is losing £10 per customer. Or at least that is the loss if customers don't repeat-buy.

If the customer buys more over the course of the year, then you're making money from the second sale onward, but until then, you have the cost to acquire the customer and that cost takes time to recoup. Plus, it could be several months from the first order before the customer uses the platform again.

You can see why marketplace businesses have strong time-to-money constraints and often need a good amount of starting cash to get off the ground.

This marketplace example illustrates why the time-to-money profile is so important. Even when your idea is progressing well, it can take a lot of time to establish your idea and meet your objectives.

To put this time-to-money position of marketplaces into perspective, here is a quote from a venture capital firm that sets expectations about how long this type of business takes to become successful: [132]

> *Considering you need to establish both buyer and seller communities,*
> *(for marketplaces) you will need more time to prove your business.*
> *It can take three years for a marketplace to get going.*
> *– Version One guide to marketplaces* [133]

That's right, some marketplace businesses can take up to three years before the time-to-money position becomes stable. You should take time to reflect on the nature of the business model and the impact your chosen approach has on your time-to-money.

Having a long time-to-money isn't inherently a good or a bad thing — it's just something you need to be aware of and plan for. Make sure you have a strategy to bridge the gap between your launch date and time-to-money.

The model booking marketplace is an example with a long time-to-money, so here is an example of a business model that has a shorter time-to-money.

Picture a web app that helps small businesses, such as training providers, implement a process within their business to apply for tenders or bids in America. The platform helps users find tender opportunities suitable for their business, then guides them through prioritising and completing their submission. They can also use the platform to find work and see how to bid for that work.

As it's a software tool that people will use every day, you can charge customers a monthly or annual fee. This billing approach means that if the service provides enough value to the customer, they'll continue to renew their subscription each month, providing the business with a reliable recurring source of income from launch day.

This is an example of a how most Software as a Service businesses charge for their platforms. Recurring monthly payments generally enable a much faster time-to-money position compared to most marketplaces which take a cut of each sale.

Before you jump in and begin your project, you should reflect on the cash flow position of your chosen strategy. You will need a plan for resourcing any development activities. If you aim to generate revenue from your tech idea, then your time-to-money position will be influenced by the revenue model you picked earlier "Chapter 3: Ways to Make Money (Revenue Models)".

Reflecting on the types of innovation discussed in "Chapter 1: Find the Winning Idea", if your provide a product or service similar to your competition then you can change your business model to innovate and achieve success. This means you might have to weigh up the cash flow benefits of one strategy against the competitive advantages of another. If you are well-funded, you may be able to differentiate in ways that less well-funded competitors cannot.

How to fund your project

You will need to explore ways to raise additional funds to if you have a time-to-money problem for your saleable product. Or you will need to pay development costs if your idea is an internal tool that won't directly generate revenue.

I must emphasise that executing your tech idea is often a high-risk, high-reward strategy. There is a chance that you could make a lot of money or add a lot of value, but there's also a risk that if your idea fails to gain traction, you could lose all of your initial investment. So, with all of these funding options, remember that your capital is at risk, make sure you know what you're getting yourself into, and only take risks you can afford to take. I don't endorse any strategy for raising money over another. The path you take will depend on the specific culture, objectives, and environment within your business.

Here are some common ways I've seen clients I've worked with fund their projects other than from existing cash reserves.

Method 1: Loans and overdrafts

If your time-to-money is relatively short, within a year or two, you might consider borrowing money to cover the gap. [134]

The injection of cash from a start-up loan might be enough to help you bridge the gap between launching your idea and making money from it. However, be careful as you'll begin paying back your loan immediately and this subtracts an expense from your cash flow each month. For example, if you expect to need to cover a £25,000 shortfall in your cash flow, then a £25,000 loan might not be enough.

Example A

Look at this simple cash flow forecast where each column represents a new month. Each month has income from sales and outgoings from expenses.

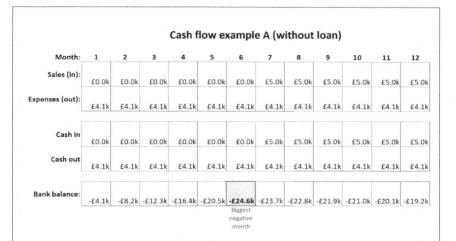

Cash flow example A (without loan)

Month:	1	2	3	4	5	6	7	8	9	10	11	12
Sales (in):	£0.0k	£0.0k	£0.0k	£0.0k	£0.0k	£0.0k	£5.0k	£5.0k	£5.0k	£5.0k	£5.0k	£5.0k
Expenses (out):	£4.1k	£4.1k	£4.1k	£4.1k	£4.1k	£4.1k	£4.1k	£4.1k	£4.1k	£4.1k	£4.1k	£4.1k
Cash in	£0.0k	£0.0k	£0.0k	£0.0k	£0.0k	£0.0k	£5.0k	£5.0k	£5.0k	£5.0k	£5.0k	£5.0k
Cash out	£4.1k	£4.1k	£4.1k	£4.1k	£4.1k	£4.1k	£4.1k	£4.1k	£4.1k	£4.1k	£4.1k	£4.1k
Bank balance:	-£4.1k	-£8.2k	-£12.3k	-£16.4k	-£20.5k	**-£24.6k**	-£23.7k	-£22.8k	-£21.9k	-£21.0k	-£20.1k	-£19.2k

Biggest negative month

You can see that the business has £4,100 of expenses every month, which sees a negative cash flow position for the first six months as the business has no income yet.

The bank balance reduces steadily each month into debt reaching a peak of minus £24,600 in month 6.

Things finally get better from month 7 onwards as the business moves into a positive cash flow position as it begins making sales, which reduces the negative bank balance from this month onwards.

As highlighted a moment ago, with a projected maximum negative bank balance of £24,600, you might draw the conclusion that a £25,000 loan is enough to bridge the gap, ensuring that the bank balance never drops below zero. But this isn't the case, and the business would actually need a bit more than that to account for the repayment schedule.

Example B

To show you what I mean, let's assume you raise just a £25,000 loan to cover the projected negative bank balance at month 6.

To do this, I've added two new rows to this version of the cash flow forecast: an income row to represent the £25,000 loan payment from the bank at the start of month 1, and an expenses row to represent the payments to the bank at the end of each month.

To keep things simple, I've assumed that the loan has no interest and will be repaid evenly over two years (24 months), which works out at approximately £1,042 to be paid back each month.

Cash flow example B (with £25k loan)

Month:	1	2	3	4	5	6	7	8	9	10	11	12
Sales (in):	£0.0k	£0.0k	£0.0k	£0.0k	£0.0k	£0.0k	£5.0k	£5.0k	£5.0k	£5.0k	£5.0k	£5.0k
Expenses (out):	£4.1k	£4.1k	£4.1k	£4.1k	£4.1k	£4.1k	£4.1k	£4.1k	£4.1k	£4.1k	£4.1k	£4.1k
Loan (in):	£25.0k											
Loan repayments (out):	£1.0k	£1.0k	£1.0k	£1.0k	£1.0k	£1.0k	£1.0k	£1.0k	£1.0k	£1.0k	£1.0k	£1.0k
Cash in	£25.0k	£0.0k	£0.0k	£0.0k	£0.0k	£0.0k	£5.0k	£5.0k	£5.0k	£5.0k	£5.0k	£5.0k
Cash out	£5.1k	£5.1k	£5.1k	£5.1k	£5.1k	£5.1k	£5.1k	£5.1k	£5.1k	£5.1k	£5.1k	£5.1k
Bank balance:	£19.9k	£14.7k	£9.6k	£4.4k	-£0.7k	-£5.9k	-£6.0k	-£6.1k	-£6.3k	-£6.4k	-£6.6k	**-£6.7k**

Still negative balance

Even with a £25k loan, the bank balance will still become negative in month 5 and stay negative until the end of the year where it grows to minus £6700.

The business had to begin repaying the loan in instalments from month 1 onwards, which means that the £25k loan was not enough. The cash flow forecast shows that with the current revenue plan the business will likely needed to borrow at least £7,000 more to keep their bank balance above zero for the rest of the year.

If you consider a loan, then make sure you plan your cash flow forecast so that you can see what happens to your bank balance as the loan repayments come out each month. If the balance goes below zero, then you'll need to either acquire more funds or adjust your plans.

Overdrafts are another form of short-term loan that banks offer to help businesses manage their cash flow. Rather than borrowing a fixed sum, an overdraft is a pre-previsioned amount that your bank balance is allowed to be negative. For example, if a bank allowed a £5,000 overdraft, they would allow you to have a temporary negative bank balance of up to that amount. They will then charge you interest payments based on how deep you go into your overdraft and for how long.

The flexibility of overdrafts makes them a popular option for many companies and they are used in combination with other forms of finance including loans.

Method 2: Business partners

A business partner is someone who owns a sizable stake in your business and works in your business at a similar level of seniority as you. 50/50 partnerships are common where two founders begin a business at the same time, each making similar money or time commitments. Just under half of all start-ups have a single founder, while just over half have two or more. [135]

Going into business with another person is like marrying them but more stressful, so pick your partner(s) with care. It's usually best to explore the networks of people you already know well, such as those you've worked with in a commercial context.

You should always be careful giving away a portion of your business because once it's gone, it's gone. In my opinion, you should only partner up if that person brings as much or more value to the table than you over the medium to long term. Ideally, the value they bring should be different to the value you bring. This value can be in the form of their knowledge and experience, time, connections, infrastructure, or money.

If you're self-funding your business idea, then finding a partner willing to match the funds you're committing is one way to avoid hitting a point of negative cash flow.

If you're exploring a partnership with someone else, make sure you sit down and discuss all of the what-ifs along your journey. For example, what if one of you loses interest? What if the business runs out of money? What if the time commitments from each partner don't turn out to be balanced?

It's much better to agree on what happens in those scenarios now, before you've both committed significant time and resources. Once you agree, you should write it down in a Shareholders Agreement, which you both sign. [136]

It's common for non-technical founders to want to partner with technical ones to deliver the technology. I'll cover the pros and cons of this later in "Chapter 11: Ways to Build Your Tech Idea".

Method 3: Raise investment

Another way to resolve a negative cash flow position is to raise money from external investors. Investors put money into your business in exchange for owning a portion of it. The portion they own after investing is called their 'equity stake'.

Investors are different to partners as they expect you to do most, if not all, of the day-to-day running of the business. They want to put their money to good work, not their time.

You don't usually have to personally repay the funds an investor put in as they were buying an equity stake in the business, which is often around 5–20% depending on the amount raised and the life stage the business is at. That said, the investor *will* expect you to deliver results as they invested their money with the goal of making a return. This might be in the form of profits later or the business increasing in value, enabling them to sell their share for more money later, called an exit.

The benefit of raising money this way compared to a loan is that early-stage investors often won't expect to see a return on their investment within the first few years, which can give you ample time to grow the business and become profitable.

Investors in early-stage start-up businesses generally fall into one of the following categories:

1. **Friends and family:**

   As you might expect, these are people you know and trust who might want to help you out by giving you money to get started. This is often the easiest way to raise money, but you need to be careful that they understand the risks — should your idea fail, you might never be able to repay them. And even if you can repay them, it might take several years.

2. **Individual angel investors:**

   Angel investors are typically wealthy individuals looking to own a share in your business. Like any investor, they will want a return on their money, but often their motivation for investing is their passion for entrepreneurship. [137]

   Many governments across the world have incentives to encourage investors to risk their money in start-up businesses. For example, in the UK, there is a scheme that allows investors to invest in start-up businesses in a way that qualifies them for tax relief. [138] This scheme means that if an investor is due to pay a certain kind of 'capital gains' tax on one of their investments, such as selling a property that has appreciated in value, then investing in one of these schemes can reduce their tax liability.

3. **Angel groups (syndicates):**

   Angel groups are where several angel investors get together to pool their funds. They do this to spread their investment across several start-up businesses to reduce the risk of any one company failing.

   Government investment schemes also encourage investors to form a group, as these schemes often have limits on what single investors can invest in any one business. These limits force investors to back several companies at once to max-out their allowance.

   Angel group schemes also benefit the business founders by allowing them to raise a larger overall investment than by approaching investors individually. [139]

4. **Venture capitalists:**

   Venture Capitalists, or VCs for short, are formal investment companies looking to invest in start-ups at different stages of their business journey. VC's can typically invest much more cash than angels or an angel group. [140]

But — and it's a big but — VCs are looking for companies that have the prospect of a *very large* return on their investment. If you're looking to grow a company that will be worth £50m+, then a VC might be interested, but anything less and you'll be lucky to hear back from them.

VCs are also much stricter on what they expect from you and your business than other investors. When you raise money via this channel, you will have to agree to their stringent terms and conditions, and they will expect a lot of commitment from you. This gives VCs a great deal of power over you, and they can even fire you from running your own business if they feel you aren't performing. So, although you can raise more money this way, it has serious downsides.

5. **Crowd funding**

   Crowd funding is where a business aims to raise a large sum of money by raising small amounts of money from many investors or purchasers.

   There are generally two types of crowd-funding platforms. Some allow you to generate sales in advance of building your product or service platform. [141] Others allow you to offer a percentage of equity in your business exchange for cash from a pool of investors. [142]

   Crowd funding can be a great way to raise money as it enables you to raise small amounts of money from many people, rather than having to convince a smaller number of angel investors to part with much larger sums of money. This is arguably a market of casual investors (or purchasers), which would not exist without such platforms.

   However, this method does have some drawbacks. To be successful in raising funds from this channel, you will need a well-constructed campaign with a clear product-market fit, strong credentials, and glossy sales and marketing assets — even if your business doesn't exist yet and is still in the research and development phase. This extra groundwork means upfront costs, and there is no guarantee that your campaign will be successful. [143]

6. **Private equity (medium to large businesses only)**

   Private equity investors are similar to VCs except they are only for more established business. Private equity funds tend to invest far larger sums of money, multi-millions or tens of millions, but they expect businesses to have a proven track record and reliable revenue. So, if you're a new start-up, this option won't be available to you until much later in your journey if you achieve success.

   However, if you are an existing business that's been trading a while and has more than a million or two in revenue (dollars or sterling), then private equity can be an option, especially if you're looking to launch a new tech product that fits into your existing ecosystem of products and services or that aligns closely with your existing audience.

   This Private Equity funding method could see you raising enough to develop some serious technical capabilities, but it's a big commitment — and if you're used to having full control over your small but successful enterprise, then it may not align with your values to have an investment company with board-level control over you.

   As with VCs, it's common for private equity contracts to include a clause that allows them to sack members of the management team or directors if they feel the person or people are performing poorly. That includes you!

   Finally, you should make sure you understand the motivations of each private equity firm. Though some are in it for the long-term dividends and want to hold a stake for the long haul, others are investing with the aim of an exit in three–five years, whether that be a larger company buying your company or you listing your firm on the stock market for a public sale (known as an IPO).

   Though private equity comes with some warnings, it's a tremendous way to raise a staggering amount of money if your existing business has a strong track record.

7. **Rights issue (large, publicly traded businesses only)**

If you're a publicly traded company, then you can create new shares in your business, then sell them to existing investors or the wider public stock market to raise additional funds.

This can be a great way to raise funds, but it's critical that the investors understand why you're raising money this way. If they aren't convinced by your reasoning, then you may risk them thinking you're raising money defensively because you're in trouble rather than to fund your growth aspirations, and this might negatively impact your company's share price. This is because it's common for large companies to raise money this way when they have a strong business but an unforeseen event might put the company's finances at risk.

For example, when Covid-19 struck in March 2020, Rolls-Royce Plc, a large British company that manufactures jet engines and nuclear submarines, found itself in a market with suppressed demand for airplanes. This shock would likely pass in the long term but cause a lot of business pain in the short term. In October 2020, they triggered a rights-issue to raise £2bn ($2.75bn) to get through the crisis. [144]

If you're looking to raise investment, you will need to be able to show investors that you have a strong value proposition and that your business is viable and attractive.

To best present your business to investors, I've created a template set of presentation slides with accompanying instructions that you can use as a starting point to create your pitch deck. This template is available on the Execute Your Tech Idea website. I hope this will increase your chances of raising the funds required to execute your tech idea.

The template deck contains the following example slides:

1. Intro slide
2. Your team
3. The pain

4. Your solution

5. Your unfair advantage

6. The competition

7. Your business model

8. Market potential

9. Market strategy

10. Financials

11. Funds / investment needed

12. Summary (why your business?)

Each slide contains instructions in the notes section about what to include, and you can simply fill in the blanks based on your business.

Free Resource Pack

The Execute Your Tech Idea website contains further information, quick-start document templates, and other helpful free resources

executeyourtechidea.com

Method 4: Launch in stages

As discussed earlier, it's very tempting to come up with the perfect vision that encapsulates every feature you could ever imagine. The lessons in our MVP chapter taught us that the problem with this strategy is that it creates waste and delays your launch. Building the perfect idea takes time, and every month that goes by where you're still 'in development' is a month where you aren't selling and making money.

Most large-scale businesses didn't start out delivering the products and services that they offer today right from the beginning. They underwent a process of evolution, beginning in one market with one strategy and growing out from there. This evolution allowed them to compete as their resources allowed whilst their enterprise was smaller.

Jeff Bezos started Amazon by selling books online before evolving the business model to offer a much wider range of products and services. Elon Musk began by launching a luxury high-end sports car to prove the business model and build innovations before entering the mid-range car market with the Model 3. Neither of these businesses would be as successful today if they had tried to deliver their endgame at day one.

When you launch a high-tech business, you will encounter a phenomenon called "crossing the chasm". [145] This is where the effort to get early innovators and early adopters to use your product is around a factor of ten times less than achieving mainstream adoption. These early adopters are generally more willing to try something new and pay for the privilege of doing so, but this behaviour is not translated into the wider market of buyers you will need to reach if you want to hit mass adoption.

Though crossing the chasm will be a future challenge for your business to overcome, it also represents an opportunity for your business to recognise the power of reach and prove your ideas to those early adopters.

Think back to the MVP philosophy discussed in Chapter 7: Your Minimum Viable Product (MVP). Another way to approach launching in stages is to again ask yourself: What is the next smallest thing I could build and convince an early adopter to pay for? Even if the look and feel of that first thing is a significant departure from your end vision.

Think of this approach as your minimum viable *next iteration* rather than minimum viable *product*.

For example, imagine your business was to create an online portal to track IoT devices, such as smart lightbulbs. You may have grand plans to launch an all-singing-all-dancing cloud platform that enables the remote management of thousands of interconnected lightbulbs, and you're confident you could sell recurring monthly subscriptions to large businesses and hotel chains.

These big businesses are the mass market customer beyond the chasm, and they will need you to convince them that your product has credibility to deliver on its promises. This means your product will need to reach a significant stage of maturity, which will result in sizable design and development costs before your business is ready to aggressively launch and sustain a position in this market.

Instead, maybe you could start by creating a simple consumer product, with a per-unit price, giving consumers the ability to connect to their smart lightbulbs directly if they are connected to their home Wi-Fi router. This eliminates the need for and cost of building a comprehensive online portal at the beginning.

Once this initial feature is built to allow you to make money, you can focus on your next Minimum Viable Iteration. And if that iteration adds enough additional value, you can update your prices to include it as an optional extra.

This staged Minimum Viable Iteration approach gives you a much faster path to revenue, allows you to target the early adopters who might be happy to pay more to play with new technology, and begins to build a foundation brand that you can expand on to become the larger, more feature-rich platform you envisioned at the beginning.

Method 5: Change business model

Think back to the "Chapter 3: Ways to Make Money (Revenue Models)" chapter earlier. Changing your business model is another way to solve a short-term cash flow problem, allowing you to capitalise on your business and fuel growth by re-investing existing revenues.

For example, if you aim to sell your product or service as a monthly subscription, this is affordable for the end user, and you can sell the benefits of not being tied into a fixed contract. However, as payments arrive monthly, you don't get the full financial benefit of acquiring a customer until they've been a customer for some time. This is a common problem among Software as a Service (SaaS) start-ups, which tend to opt for a monthly billing model.

To combat this cash flow problem, you could launch your business with annual rather than monthly subscriptions. This is where you advertise your pricing based on what your offering is worth per month but require new customers to buy a one-year subscription in advance. This approach creates a higher barrier to entry in signing up new paying customers as it's a bigger commitment, but you get a full year of fees from them upfront, helping you to eliminate your cash flow problems.

Even if you plan to switch back to offering a monthly payment option later, you can keep the annual plan as a discounted option. Customers benefit because they spend less, but you benefit because you get a year's worth of money upfront.

So if you see a time-to-money problem, reflect on the different business models that exist, and see whether the time-to-money position changes for one compared to another.

Closing words

It's possible to be successful with your idea if your time-to-money is long, short, or somewhere in between. There is no right or wrong answer. However, if you don't manage your cash flow and plan for how long it will take to make sustainable revenue, then you are doomed to fail.

Give yourself the best chance of success and plan your finances. If it helps, use a cash flow forecast to predict if and when your bank balance will go into negative numbers and by how much, then use that information to guide your funding decisions.

Don't be afraid to revise your idea or approach so that it better fits your cash flow or funding situation — it's just part of doing business.

Key takeaways:

1. Some projects can become self-sufficient, supporting their ongoing development or maintenance costs through the revenue they generate.

2. Some projects add value indirectly and must be paid for using your existing funds or by raising money.

3. If you have a shortfall in your existing funds or a time-to-money cash flow issue, then you may need to consider raising additional money to support your project.

4. If you aim to make money from your tech idea, then reflect on the various ways of doing that from "Chapter 3: Ways to Make Money (Revenue Models)" and model how they impact your time-to-money position.

5. Loans, partnering, raising investment, launching in stages, or changing business models are common ways to address a shortfall in funding.

6. Don't be afraid to change your approach to make your cash flow situation work.

CHAPTER 9:
HOW TO CONVINCE KEY PEOPLE

"Leaders must earn the trust of their teams, their organizations, and their stakeholders before attempting to engage their support."
— *Warren G. Bennis*

Chapter Relevance	
New Internal Tool	**New Product**
↑	↑

In the last chapter, we discussed the different ways you can raise money to get your idea off the ground. But what if your ability to get started relies on more than just securing the funds? What if you also need to make a compelling business case for your idea to those around you, such as banks, managers, directors, or in some cases even your family? This chapter offers my recommended approach to convince people in your team, start-up, or organisation that your plans for building a tech idea are worth their time and money.

If you're building a tech product within an existing organisation — particularly one with an existing internal tech team — you're going to need buy-in from internal stakeholders to approve the budget for the project to go ahead. But it's not just about the money, and the greatest barrier facing innovation teams is a lack of commitment on the part of top managers and low involvement and participation by team members. [146]

If you're a start-up, unless you're a one-man-band founder, you will require stakeholder buy-in too. If you're a non-technical founder, then you may need to make a convincing argument if you want a new set of features — or even to prove that your vision for the direction of the tech is worth pursuing.

No company will let you use their budget to build a product without reviewing it first, unless you've gained an incredible level of trust. So, you're going to have to convince the relevant people that they should approve or support your idea. Everything you've learnt about placing a value on your product so far wasn't just relevant to prioritising what you will build — it's also a crucial part of getting stakeholder buy-in.

The stakeholders you need to convince will want to see evidence that you've thought through your plans, and you need to present them in a way that boosts your credibility.

It's wise to put together a brief proposal or business case before seeking buy-in. This can be informal and may be a few slides or a written document depending on your company's processes. Explain the problem that your product will solve, how the development will be managed, and what help you might need from the internal IT team, external resources, or other stakeholders or teams in your business.

Here are the main questions you will need to address before you speak to the key decision-makers. The more quantitative you can be in your approach (making your case objectively with numbers, data or evidence), the better!

1. How does the project relate to your objectives, the department's objectives, and therefore the objectives of the business?

Your idea will need to align with the identity and values of your organisation, your departmental objectives, and personal objectives within that department. If you have a mismatch in values or objectives, it's unlikely that you will gain approval to execute your idea.

Consider our previous Apple example — their whole business model is based on giving status to its customers through high-value, high-quality products. [147] However, the market for low-cost devices is huge. If you were working for Apple and spotted an opportunity to launch a low-quality but low-cost product to meet the demands of that market, then your idea would conflict with the company's vision and values. This conflict makes your low-cost product idea very unlikely to gain approval in that setting.

If you can remind key stakeholders what the organisation says it stands for, then align your idea and its value to those values, then you will increase your chances of success.

2. What resources are needed to fulfil the project and where will they come from?

Whether it's infrastructure, money, or people, every resource in your business will be owned by someone who is responsible for its effective use. If people are responsible for something, then they are protective of that thing. Why should they give you access to their valuable resources if it might conflict with their other goals?

You will need to think hard and be honest about all of the resource commitments required to deliver your idea. List every single one, how much of it you need, when you need it, who has responsibility for it, and what the cost is to the business. Every person you list who is responsible for resources or money you need must be involved in the decision-making process — and you will have to convince them that your idea is worth the risk.

3. Why do you need the money or resources and is it worth it?

You'll need to demonstrate the financial worth of your product, pinpointing how much you expect it to cost along and how much the company has to gain.

For example, it might cost £50,000 to develop your tech idea but will save the IT team £150,000 over the next three years, which justifies the initial cost. Most companies will want a picture of the next 3 years of costs: what it will cost to develop, what it will cost to maintain, and how much the inevitable phase 2 of development will be.

Imagine you're the head of IT or the technical director. Your teams are busy and someone from another department needs your support on a project that will keep one of your team members occupied for six months. You also need to oversee the project, which risks distracting you from projects from other departments, which are already in the pipeline. Each project has deadlines you're committed to, and that are important to deliver results for the business.

The Opportunity Cost concept we discussed earlier applies here too. The stakeholder you wish to influence might need to consider the opportunity cost of assigning resources to complete your project over someone else's.

In this scenario, you need to be certain that the new project is worth the effort and that other key people in the organisation are likely to support it.

4. Where will the money come from?

For small businesses, there will be one person who owns the whole budget for the company, but for larger organisations, it's common for each department to have its own budget, and a manager who is responsible for spending that budget. Although there are pressures to spend the available budget efficiently, the good news is that there are also pressures to spend *all* of that budget.

If senior management assigns a budget to a department and that department doesn't spend it, it's either a sign that management isn't doing enough to make the most of the money (and there is a performance problem) or that the department has become more efficient and no longer needs the money. Even though the later example may seem desirable to the business, in practice no department head wants to see their budget cut!

For technical projects, it isn't always clear which budget the project should belong to. For example, if it's an IT project, should it use the IT budget? What if you're building a tech product — maybe the budget should belong with the product development department? The project will deliver new client leads, so maybe the budget should come from marketing instead?

As you can see, this may get a little confusing.

Given the nature of your idea, you should be able to come up with a rough idea about which departments it influences. Have a conversation with each department to determine whether they feel that elements of the project could theoretically be covered partly or fully by their departmental budget. The culture of your organisation will influence where the money comes from and whether you're able to combine budgets from different departments to fund your idea.

5. Do other key stakeholders support the project?

If the department you work for is willing and able to cover the project cost using their budget, then project approval might be simple. As discussed in the last point, the decision will often straddle several departments and competing interests. You'll probably have to involve your department, IT, and finance as a minimum, but other departments such as marketing may need to be involved.

It's unavoidable that you must think about internal politics. You don't want to be seen as the child who didn't get the answer they wanted from mum so went and asked dad instead. Proceed with care.

There are three main types of power that influence decisions in organisations: [148]

Role power: This where your position or authority in the organisation means you can push through an idea because of that position. This is a blunt weapon, and you should avoid using role power if possible as it will often erode your relationships and ability to be effective.

Expertise power: This where you have unique skills or knowledge that give your opinions and instructions more authority. You will give more weight to marketing recommendations from the marketing expert than a non marketing expert, such as the accounting team.

Relationship power: This where you've built connections and relationships with various people in the organisation and they want to help you because there is mutual like and respect. This is the most effective way to influence decision making, and I think it's also the most rewarding. Who wants to bulldoze their team to adopt ideas against their will? It's much better to move forward together, listening to each other and acting out of shared interest and mutual respect. Building a culture where people act this way also results in happier teams and better staff retention.

Once you've identified everyone who is likely to have an interest in or objection to your idea, take the time to build relationships with them. Find reasons to talk to them, or better yet, arrange an informal one-to-one with them, such as a working lunch. Once you've built a relationship, you can raise elements of your idea and explain your thinking gradually, seeing what they think and if they have any objections.

If the stakeholders you need to convince like you and know a little bit about your ideas, then you are ready to make your move and formally push for project approval.

6. What is the project's timeframe and the ongoing commitment of time or money once it's launched?

Setting and committing to action is an important part of offering an effective, professional project, and other stakeholders will expect you to know the project timeframes and any other ongoing commitments of resources or cost.

Learn what is needed to plan and deliver the project, both from your company and any suppliers who will support its delivery. If the project is too early to commit to firm timeframes and costs, then work with ballpark estimates based on your best assumptions. You don't need to account for every cost, deliverable, and day to achieve a broad buy-in for the vision.

Closing words

To help with this task, I've created a set of slides, available on the Execute Your Tech website, which show you how to prepare all of this information to present to the key stakeholders objectively and quantitatively. These slides have the following sections:

1. The opportunity

2. Valuing the idea in financial terms

3. Value to the business

4. Objection audit

5. Resources needed

6. Return on investment

7. Timeline and next steps

8. The summary

On each slide, look out for the notes section (you may need to turn this on in PowerPoint), which give further guidance on what to write and how to prepare each section.

Free Resource Pack

The Execute Your Tech Idea website contains further information, quick-start document templates, and other helpful free resources

executeyourtechidea.com

Be careful not to assume that just because your project requires IT expertise to manage, it will be owned by or maintained by your IT department. IT often has limited scope and resources, doesn't have the capability or capacity to develop products or solutions, and may not have developers. In these cases, you'll have to source outside suppliers and account for the cost and process of doing that.

Make sure you know your company's processes for bringing in third-party help. Also ensure you know whether your product will have to be made by a team of pre-approved suppliers or whether you have free reign to choose as long as you stick to pre-approved criteria. If your company has a large IT department with in-house developers, they might be capable of carrying out the project themselves. In these cases, you may only have to win over the development team.

Speak to key people one by one to gain allies and patiently win buy-in for your ideas. Once the most important people are on-side, you can then release your proposed ideas more widely with lower risk of rejection.

Key takeaways:

1. Unless you are a sole-founder start-up, you will most likely need to convince other people or stakeholders that your project is worthwhile.

2. Internal money and resources may seem abstract but they have a real cost. Remember that your project has an opportunity cost impact to others in the organisation. Consider and handle the objections they might have in advance to increase the chances of your project being accepted.

3. Take time to audit and understand the interests of any stakeholders who are pivotal to your tech idea.

4. It's important that you understand your expertise power, role power, and relationship power before attempting to convince key stakeholders.

5. The slide pack provided with this book contains a way for you to communicate your ideas and value proposition to the stakeholders you want to convince.

CHAPTER 10:
DOCUMENT BEFORE YOU IMPLEMENT

"Project management can be defined as a way of developing structure in a complex project, where the independent variables of time, cost, resources and human behaviour come together."
— Rory Burke

Chapter Relevance	
New Internal Tool	**New Product**
↑	↑

By this point, you should have identified the problem you're aiming to solve and a solution to that problem. You've taken steps to validate your idea, making sure it's both commercially feasible and adds value to the end customer, user, or your organisation. You've also used an impact-effort matrix to determine what your priorities are.

Now it's time to get specific

This chapter will guide you on simple ways to write or present your ideas so they can be communicated to other people, which will be useful if you need to commission external suppliers to build the project for you.

It's time to put together a plan of action, so you need a scope and a requirements specification. This chapter details what each of these is, what stage they are needed, and who should create them.

Imagine we're in a world where cars don't exist yet, and you're an inventor who has just thought up the concept of a car for the first time. In this hypothetical situation, you know you want a black car with four wheels but don't know what is involved to bring your vision to reality.

In practice, making a fully functioning car will require you to figure out how every intricate detail of the engine should work so you can instruct an engineer to meet those detailed requirements.

This full, detailed plan is what we call a technical specification, and it provides all of the details about all of the parts of the car and how they should all work and fit together.

However, you aren't an engineer, and you don't know how to build a car, so how on earth are you supposed to draw up a technical specification that an engineer can follow to build one? Sure, you can tell someone you want a black vehicle with four wheels and an engine that consumes fuel to make the wheels turn, but you don't have the expertise to give specific instructions so they can engineer each individual component into one.

This illustrates why it's unreasonable to expect non technical people to write highly technical instructions. If I expected this of you, this book would be called *Executing Tech Ideas for Hardcore Techies*!

Even though you aren't expected to give highly technical instructions, you do need to get your ideas down on paper in a meaningful way so you can properly delegate the process, either to your internal tech team or an outsourced agency (which we'll cover in more detail in the next chapter).

The technical instructions that outline the specifics of how your project must be built are usually called the specification, though sometimes people use this term to mean different things. There are three types of documents that I've heard our clients call a specification:

1. **The scope:**

    A simple, non-technical document outlining what you want.

2. **The non-functional specification:**

    The specifics of what you want your project to achieve without saying how it should achieve it.

3. **The technical or functional specification:**

    The specific details of exactly how the project should deliver the non-functional requirements.

You can write your own scope but should delegate writing the technical specification to your chosen tech team.

I'll expand on the purpose of each of these specifications:

Type 1: The scope

Your scope is a thin version of your requirements. It references the high-level objectives of the project but doesn't usually go into the finer details of how these objectives should be achieved. Most scopes I've seen are one or two A4 pages of bulleted notes with a background section at the top that provides an overview of the project.

The purpose of the scope is to ensure that stakeholders working on the project have enough information to engage in future discussions and planning activities. Your scope is a foundation to be built upon. And it's normal for this foundation to adapt and change as you progress through the stages of planning.

I'd argue that changing your scope early-on is desirable and it's fine to revise it a few times as you change your mind about ideas. This is because every change you make in planning can save you thousands of pounds compared to making those changes mid-way through design and development. Early changes are cost-effective changes.

A simple scope will often contain the following sections:

1. **Background**

   A brief description of the company, the problem to solve, and a couple of sentences describing what the project aims to achieve. This is sometimes called the project scope description.

2. **Project goals and achievements**

   An itemised list of what you would like the project to do. This might be to reduce admin time or enable some previously impossible process.

3. **Target milestones and budget**

   A rough outline of what the business aims to achieve, when, and for what cost. It's fine for this information to be approximate at this stage.

4. **What features/requirements are necessary**

   This is a bulleted list of everything you believe is required in your project to meet your objectives. For example: user login accounts, admin portal for administering data, online payment system accessible to end clients, etc. You can follow the techniques from "Chapter 7: Your Minimum Viable Product (MVP)" using the impact-effort matrix to identify what is necessary.

In my experience, customers who have taken the time to write a detailed scope are much more invested in their projects success.

You can find a project scope template in the free resource pack available on our website. [149]

 **Free Resource Pack**

The Execute Your Tech Idea website contains further information, quick-start document templates, and other helpful free resources

executeyourtechidea.com

You can optionally extend your scope to include additional information such as key stakeholders, version history, or a more detailed success criteria.

Remember that the main value of a scope is to get your ideas down on paper in the way that works best for you. If you feel better drawing hand-drawn sketches, then do it. If you'd like to include screenshots from other apps and systems you like, do that too.

Type 2: The non-functional spec

A non-functional requirements specification is a beefed-up version of the "what features/requirements are necessary" section of your scope. It outlines what needs to be done without going into the technical detail of how it can be done.

For example, if you want to build a feature for managing users, then the non-functional requirements might look something like this:

Example

1. *Deliver functionality to allow the creation and management of users.*

   a. *Administrators should be able to add new users.*

   b. *Non-admin accounts should not have access to this section.*

   c. *A facility should exist to allow for admin users to search for users within their organisation.*

Notice how this example says what needs to be achieved and is a lot more specific about which individual features are required when compared to the scope?

Writing a non-technical specification is a step-up in detail from writing a scope and will take more time. Unless you are a technical person with experience building systems, you will probably overlook features that you need or make incorrect assumptions.

It's fine and completely normal if you don't feel technically minded enough to write the technical specification. It's usually something you'll delegate or work collaboratively with your technical team to produce.

Type 3: The functional specification (technical specification)

The technical specification is the longest document and covers in detail how each function should work. Remember our user log-in example from earlier. Here is an example of how we can expand on the requirements further in the technical spec:

Example

1. *Deliver functionality to allow the creation and management of users.*

    1.1. *Administrators should be able to add new users.*

    1.1.1. *The interface should display an "add users" button.*

    1.1.2. *When the "add users" button is clicked, it should take the user to the add users screen.*

    1.1.3. *The add users screen should contain the following input fields:*

    1.1.3.1. *User email (validated as an email address, required).*

    1.1.3.2. *User password (string, required).*

    1.1.3.3. *Enable two-factor authentication (Boolean).*

    1.2. *Non-admin accounts should not have access to this section.*

    1.2.1. *If a non-admin user visits the URL path to this page, they should be greeted with a "You do not have permission to access this section" error message.*

    1.2.2. *Admin users should have full access to this section and its functionality.*

    1.2.3. *User permissions should be delivered via the permissions system outlined in section X.X.XX of this specification.*

1.3. A facility should exist to allow admin users to search for users within their organisation.

1.3.1. The search bar should be hidden unless the search icon is clicked.

1.3.2. The search icon should expand to show an advanced search area.

1.3.3. The advanced search area should contain the following fields:

1.3.3.1. Search by email.

1.3.3.1.1. Performs an SQL "%LIKE%" search on the user.email database column.

1.3.3.2. Search by date.

1.3.3.2.1. The user is presented with a from and to date.

1.3.3.2.2. When the search is conducted, all users who were added or edited within the from or the to date should appear in the results below.

1.3.4. Search results should appear in a tabulated results area underneath the advanced search box.

1.3.4.1. The results should be paginated, displaying 20 results per page.

1.3.4.2. The following table columns should display on the results data:

1.3.4.2.1. User email.

1.3.4.2.2. Date added.

1.3.4.2.3. Actions.

1.3.4.2.3.1. Edit button.

1.3.4.2.3.2. View button.

1.3.4.2.3.3. Delete button.

You can see this specification includes significantly more detail when compared to the scope and non-functional spec. What was four-lines of explanation is now expanded to 27!

You will also notice a bit of jargon in there too, such as references to database query types and validation rules. This language is unavoidable in such technical documents and demonstrates the technical knowledge required to complete this spec.

As with the other idea and validation phases, these three document types are evolutionary. You will begin with your scope, then expand it to create the non-functional spec, and expand that to create the technical spec. Each time you revise your specification documents, you can mould and change your ideas as you problem-solve with your technical team.

Bonus type: Wireframes

We already covered the basic benefits of wireframes as a prototype in "Chapter 7: Your Minimum Viable Product (MVP)" but I'll elaborate here.

Wireframes are sketches of how a user interface should look. Wireframes aren't just useful as a prototype to validate your MVP — they are also critical to support your written specification. Some things are more easily communicated with pictures than words, and without wireframes, you risk too much being left to interpretation.

For example, close your eyes and imagine you need to give me instructions on how you'd like me to paint a black spotted dog. [150]

- Is the dog tall or short?

- Is it a dalmatian breed or a Beagle?

- What pose is it in?

- Is it a black and white or colour painting?

- Is the dog on its own or part of a scene?

There is a lot of detail to discuss to ensure the brief is met correctly. I could write an essay describing every aspect of the dog, which even then I'm sure we would interpret differently, and it might result in an end product that isn't what you wanted. Or you could show me a picture to remove the confusion about what kind of dog you imagined.

Wireframes should show the structure and positioning of the components to highlight the overall concept, without the graphic designs applied.

This is what a wireframe might look like for a hypothetical section of a new mobile app project:

Hand-drawn wireframe   **Digital wireframe**

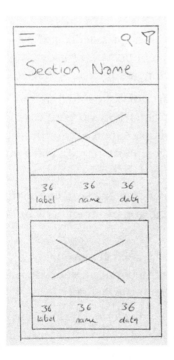 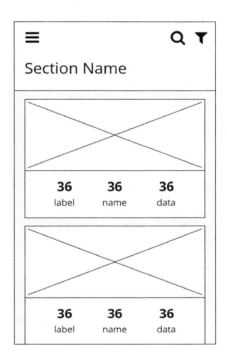

If you're a visually creative person, wireframes enable you to experiment with different layout options and communicate your ideas quickly and effectively. Many people find wireframing a fun and rewarding task, and you may enjoy having a go yourself!

However, if you do create your own wireframes, I still highly recommend having them professionally made by your chosen tech team to accompany the technical specification. [151] You should make the most of the fact that your chosen tech team will have more experience designing interfaces and user experiences than you. This doesn't mean that any wireframes you made are wasted though as they're a great tool for communicating your ideas visually to ensure the team understands what you want.

Here are some quick-start tips for creating your own wireframe sketches:

1. Tooling doesn't matter too much, and wireframes can be super rough, so it's fine to sketch them on paper with a pencil, use shapes in PowerPoint or other spreadsheet software, or use specialist rapid-wireframing software. [152]

2. Start by drawing a square that matches the proportions of the screen you want to wireframe. This might be a portrait rectangle for mobile devices or a landscape rectangle for desktop devices.

3. Draw rectangles in the area to split the interface into different logical components. Then either describe what you want each section to be in words, which is the simpler approach, or draw the interface elements you want to see, which may require you a little more design experience.

4. Get inspiration from other interfaces or apps that you like or that have similar requirements to yours, and you can then draw or label each section.

5. Try to keep each wireframed interface screen consistent. For example, if you've used a menu at the top of one screen, you should probably continue to use that convention throughout.

6. Your wireframes should be functional, not pretty. The aim is to communicate rough ideas that are allowed to change later in the process, so you don't need them to be too good.

7. Add comments to your wireframes where the sections you've laid out need further explanation. For example, if you've included a button, then you may want to describe what that button does when clicked.

Wireframing is a useful tool for all stages of your journey if your tech idea requires people to interact with a user interface.

Closing words

Once you're clear what your MVP should contain, you are ready to create your supporting documents: begin a simple scope document where each bullet point is a high-level requirement of your project, then elaborate to describe your non-functional requirements. Optionally, supplement these documents with wireframes that visually explain your ideas.

There are several philosophies that guide how tech projects should be planned and managed, and each type requires different forms of documentation. For example, the "Waterfall" style of project management involves creating detailed technical and non-technical specification documents before the project begins. By contrast, "Agile" methods. of project management see the technical details outlined as the project progresses. Both of these approaches have different trade-offs, which we will cover in more detail in "Chapter 13: Project Lifecycle Types" and "Chapter 17: Plan Your Product Roadmap".

Key takeaways:

1. The technical specification is a complex document that is difficult to write if you aren't an engineer or trained project manager.

2. You can write a smaller specification document called a scope to outline your ideas without getting bogged down in the technicalities.

3. Write a scope document before you engage your team or external providers to build your project.

4. It usually makes sense for the person or team who you want to build your tech project to create and write the technical specification.

5. If your idea includes a user interface of some form, you can use wireframes to communicate your ideas with pictures to remove ambiguities and avoid misunderstandings.

CHAPTER 11:
WAYS TO BUILD YOUR TECH IDEA

"Ideas are yesterday; execution is today, and excellence will see you into tomorrow."
— *Julian Hall*

Chapter Relevance	
New Internal Tool	**New Product**
↑	↑

It's time to take your scope and start putting your plans into motion.

This chapter explores the different ways to approach building your tech idea. We'll look at seven of the most popular ways to approach building tech ideas and discuss the merits and trade trade-offs of each.

The foundations of any tech idea begin with design and coding activities. If you aren't a technical person, then you might not know much about code, but don't worry — this book assumes that you have zero programming skills, and indeed many people creating apps and tech businesses today don't!

We'll go through the various approaches you can take to get your idea done. In other words, how you'll carry out coding execution as someone who is just getting into the tech world. We'll review the pros and cons of each option, and look at how to proceed once you find an approach that you think will work for you.

There are seven options for how to build your tech idea:

Option 1: You code it up

Learning to code is the most cost-effective way to develop your idea but it does cost time. You will need to invest in yourself, and it may take several years before your skills enable you to self-build a high-quality tech idea.

The code-it-yourself option is desirable if you have the freedom of time but the constraint of money, which may apply to young people or those still in education.

Coding is a fantastic skill to learn, and even if your idea fails. Acquiring coding skills changes the way you think and will stick with you for the rest of your life, forever influencing your decision-making.

Even if you don't plan on coding your idea yourself, it can be quick to learn the basic principles and make you more effective at communicating with your technical team.

There are some fantastic online resources to help you learn how to code such as Code Academy, [153] Code with Mosh, [154] Khan Academy, [155] and Reddit. [156] A quick online search will reveal many more.

Option 2: Do it Yourself (DIY) platforms

Some platforms enable you to take a DIY approach to building processes and enable you to make functioning tools without needing to know how to code. The nature of your idea or the sector you operate in may determine the type of tools available to you, and there is a wide continuum of these tools.

For example, earlier in the book, I recommended you use a form-building tool like Google Forms to construct a prototype product. If your tech idea is as simple as a few web forms, then it may be unnecessary to develop your idea beyond that stage, and you can continue using the forms you've already made.

The benefit of DIY tools is that if your requirements are simple, the time it takes you to implement them can be faster than coding a bespoke solution. If you're a relatively technical person, then a DIY approach allows you to tinker with the tool yourself, and you may discover capabilities that you hadn't thought of and can bring immediate value.

The downside of DIY platforms is that their limitations might hold you back. It's very common to have great momentum in developing your ideas up to a point before hitting a brick wall where the DIY tools don't do what you need them to do. Getting past these walls sometimes requires you to start from scratch with custom programming work. So although a DIY platform can be a great way to test your early assumptions to create your MVP, be mindful that a year or so in, you may need to replace what you've made with your own custom code. [157]

Another thing to consider with DIY tools is that they often offer a per-user pricing model. This is good for internal tools where the number of users is low, but if you anticipate a high number of users, then the ongoing monthly cost may grow so high that it's cheaper to build your own app in the first place.

The final risk is that you don't own the platform, and if the DIY platform goes bust, then so does the platform you've worked hard to create. This happened in 2016 when the DIY tool called Parse shut down, leaving thousands of developers to migrate their data elsewhere. [158] The platform was eventually bought by Facebook, who closed it for good in 2017. [159]

If you find a DIY tool that meets most of your needs but not all of them, then hybrid approaches are also an option. This is where you build your idea using a DIY tool and fill the gaps in functionality with a bespoke solution that sits alongside it (connected using an API, as first discussed in Chapter 2). [160]

Option 3: Use your internal tech or IT team

If you work for an organisation that has an internal tech team, you will probably need to involve them before you can start your project.

I've worked with lots of different sized organisations and know first-hand that the tech or IT department's capabilities can vary greatly. Most IT departments at large companies focus exclusively on provisioning software and IT services. They may have a few developers on the team, though they are often fully allocated to integrating systems together, coupled tightly with the rest of IT.

If an organisation is structured this way, then the IT team will want to oversee your project and guide you so you don't make mistakes or compromise security or data privacy. They will want to be involved in some parts of the project, such as picking a supplier or discussing the business case, but in most cases they won't do the design, development, or project management work.

Other organisations will have an in-house development team focused entirely on developing internal projects and business process tools. These tend to be organisations with an established tech product that is core to the value the business offers.

Internal tech teams are normally composed of at least five full-time people, consisting of project management, developers, and designers. Having fewer than five people can make it difficult to create a culture of development that enables people to learn from one another, and to build and maintain good processes.

Businesses with an established tech team will also need a capable Chief Technology Officer (CTO) to ensure that this team is effective and able to make strong strategic decisions. The complexity and investment needed to establish a tech team is a big reason why development agencies exist — having a full-time dedicated team is very expensive and a big infrastructure commitment for a company.

Internal tech teams always have more project work to complete than available resources to do it, and your idea will be competing for time and resources. You will need a strong business case and plan as outlined in previous chapters to convince the business that you should be allocated a portion of this team.

Even if your company has its own tech team, you may still be expected to outsource the project to speed up delivery and avoid the tech team being distracted from developing their core product.

Option 4: Get a CTO or tech founder

The CTO is a person within your business who usually has board-level responsibility and authority and is responsible for the business' overall technology strategy. CTOs are expected to have many years of technical experience, formal training, and plenty of senior business experience. Think of them as the most senior business managers who also have a wealth of hands-on technical experience.

To small businesses and new start-ups, founders sometimes want a CTO who can advise on the technology *and* conduct programming work. It makes sense, right? Why pay for development when you can give someone a cut of the business and have them develop the solution for free?

The challenge for small businesses looking to partner with a CTO is that high-quality and experienced CTOs are in demand in the market — and if they work for you for free, they are likely sacrificing a paid position elsewhere. Again, think back to the opportunity cost principle we discussed earlier.

As a result of this demand and conflicting interests, it can be challenging to convince a CTO to work for free in exchange for a stake in your business (Equity), and they will expect you to bring a great deal of value to the table in exchange for their involvement and expertise. This might mean investing significant funds into the business, demonstrating that your idea has traction, and paying them a salary in addition to giving away equity.

You might need to develop the beginnings of your idea, convince them it is worth their time to partner with you, and make progress on building out your MVP to get them on board. If you've built a first version, launched it, and are getting some early success, it may be easier to convince a CTO to join your operation. But remember, a CTO is a strategic role, so if you hire one, then it's likely you will also need to outsource development so they can focus on the business or hire an internal tech team for them to manage.

As you can see, hiring a CTO may be useful as a business grows, especially if you're at a growth stage looking to raise millions or more in external investment. But if you're at the beginning of your journey, you should be cautious about hiring a CTO as a cheap way to build your idea for free because it likely isn't! It's a strategic choice, not a cheap one.

If you work within an existing company, you should know that the larger a business becomes, the more focused a CTO's role becomes on business strategy, and meaning they spend less time on day-to-day. Relative to your business, this shift in responsibilities means that the CTO will be unable to directly help with programming and will likely recommend that you build all or part of your idea internally or outsource to a specialist agency.

Option 5: Freelancers

Working with a freelancer usually means working with one individual (or a small freelance team, depending on your needs) who doesn't work for a larger company.

This person should be experienced in the world of tech and must be capable of offering everything you need to complete their section of your project. It's common to hire different freelancers for different phases (developers, designers, copywriters, etc.), ensuring that you have an expert in each sector.

So, you might hire a freelance developer to build your application. You'd hire one individual to do the entire job and pay them a fixed amount that you agree prior to the work, or pay them for their time if you plan to work in an Agile way with a rolling monthly budget.

You could then hire a freelance copywriter to fill your application with well-written text that enhances the user experience, a tester to go through the app and make sure it meets your requirements, a hosting company to store your database, a graphic designer to create the user interfaces, an illustrator to create the graphics that appear in each section, and so on.

You can find freelancers in various ways, including via their own websites, asking for recommendations, through social media, via

Google, or on various freelance platforms. However, please be careful — carry out checks on potential freelancers and their experience so you don't end up paying for someone who isn't up to the job. Check examples of their past work, their rates, and past reviews to get a better idea of their skill level and whether they're the right person for your product. If you like what you see, set up a phone call or in-person meeting to learn more about how you could work together.

There are plenty of benefits to taking the freelance route. You don't have to pay for an entire team, can avoid things like sick pay and holiday pay, and can hire someone who has past experience. Freelancers have often worked on a diverse range of projects, enabling you to pick the person with the specific experience you need, and they will bring knowledge from across industries, helping you develop your ideas based on what they know has or hasn't worked in the past. They're also very flexible and usually aren't stuck to a 9–5 routine. If you ask them nicely and pay them well, they're often happy to do more (paid) hours in a short timeframe and push your projects through quicker, which is a big bonus if you have a critical deadline to meet.

Daily rates are generally 20–30% less when sourcing a local freelancer compared to a small agency, but there are some downsides to consider too: You will be much more responsible for project management and ensuring the quality of the code and outcome of the end product, which will carry a significant time overhead for most tech projects. Freelancers are often not as organised as a well-oiled team, won't have as established processes, and will have some limitations in their skill set. As you haven't worked with them before, you don't know how they're going to gel with your team, and they don't have a keen investment in your business, aside from this job.

Business continuity can also be an issue with freelancers, which is your ability to continue with your business if unexpected events occur. For example, if a personal problem occurs, there's no one else there to pick up your project. So, if you do work with a freelancer, make sure you have a ready-to-go backup plan for if they disappear or get a full-time job. Make sure you have access to the source code, database, server, and any production design files for what they've built, as you don't want an unresponsive freelancer to cause a business-killing event.

Overall, freelancers can be a good option for small projects and one-off jobs where the business continuity risk is small, or if you're lacking a tech team, or even for ongoing work that you don't need a 9–5 staff member for. Just do your research before signing any contracts and make sure you hire someone who is reliable and right for the job.

Option 6: Outsource overseas

Hiring overseas freelancers is similar to hiring locally. The job description is the same: an individual who you pay to do a specific job for your business without a long-term contract. But there are some differences.

One of the main perks of hiring overseas is the price difference— depending on where you hire from, of course! Hiring from countries where the average salary is far lower than your local salary can mean you get quality work for less. For example, if you live in the UK and hire someone who lives in Indonesia, the average salary is a lot less. [161] As the cost of living in Indonesia is a lot less than the UK, paying less doesn't necessarily mean you are treating them unfairly, though you should still do your due diligence so you don't accidentally support exploitative working conditions.

Though some countries have lower average wages, others are a lot more expensive. The average salary in Switzerland is 124,000 CHF, which is around £97,000 at the time of writing, much higher than the UK or even the US average! [162] So, you can expect to pay triple the price for a Swiss freelancer compared to their UK counterpart..

Language and communication problems can also be a problem when outsourcing overseas. Even a foreign freelancer who is fluent in your native language risks misinterpreting something you ask, and this problem is exaggerated if English is their second language. Such error rates come with a real cost.

It's also unlikely you'll ever be able to meet overseas freelancers face to face, so if that's important to you, I'd suggest sticking local. The ability to collaborate is further hindered by time-zone differences. For example, if you live in the UK and plan to outsource to Shenzhen, the tech hub of China, then 9 am in the UK is already 5 pm there, which doesn't give you a big time-window for project management communication.

There are also overseas agencies. These companies are set up abroad and you can work with them on the development of your product. You've probably heard of plenty of brands who use offshore companies to create their products, for example, IBM outsource thousands of well-paying jobs to programmers in China and employ more staff in India than in the US. [163]

The main reasons for hiring offshoring agencies are to source low-cost labour and sometimes to avoid local taxes in the company's founding country. When developing tech, you may choose an offshore agency because they offer the niche skills you're looking for, skills that aren't available in your native country.

Offshore agencies usually have a small number of directors running the business and a workforce that can vary in size. However, one of the main disadvantages is that you may only find limited information on how they operate. There have been many cases where offshore companies have effectively turned out to be sweatshops — with underpaid staff being forced to work long hours in horrendous conditions.

For example, consider the 2013 collapse of the Dhaka government factory, a garment manufacturing factory that made clothes for the several high-end luxury brands. [164] The collapse killed over 1,000 workers within. The brands associated with the factory were widely criticised, and rightly so!

It's also common for template code to be sold as if it were made bespoke for the customer. If this happened to you, it would create a major business headache because of licensing and right-to-use problems.

If you're outsourcing to offshore agencies, ensure you do a thorough audit of the company you're working with and check their work standards, processes, and team structure. Also, be sure to regularly review their code. If you can, visit the company in person or send a third party to do it for you. And if you outsource overseas to save money, make sure you budget for plenty of project management time and have someone who you trust available to audit the work quality.

Option 7: Local agencies

Micro agencies

A micro agency is a tiny company of between one and nine people. They're often made up of freelancers working together but can also have permanent staff members. Micro agencies can cater to the needs of SMEs and large corporations, giving you a tiny team who will be entirely focused on your product.

Companies usually choose to work with micro agencies precisely because they're so small. They enjoy the personalised experience, getting to know the entire team by name, and having the entire team know theirs. It's also more common for the company to work on just a few projects at a time, which is appealing to clients. With less work on the go at once, their team can direct all their resources towards your product.

The downside is they might not have a broad range of resources on offer and their process maturity may be lacking, which could cause important elements such as security or quality assurance to fall through the cracks. Each employee in a micro agency is likely to have many different responsibilities compared to larger teams where it's more likely that staff will be specialists.

You may find that the staff are overstretched at some micro agencies. With less flexibility when delegating tasks, your developer might simultaneously be developing tech for multiple other companies, leading to less focus and more stress. Not all micro agencies operate like this, of course, so it pays to do your research before hiring.

Business continuity issues can also arise for small agencies as they usually haven't been around for long and might not be around for much longer. I've inherited several projects over the years from micro agencies where their main developer left or the business went bust. So, put in place plans so you can continue to work on your tech idea even if this happens.

Small agencies

This is the most common size of agency, consisting of approximately £1 million in revenue, two business owners, 12 billable employees, three non-billable employees and four full-time equivalent freelancers. [165] It's also the most common choice of agency to work with, especially when it comes to creating tech. The company isn't so huge that your product feels lost among a sea of other products, but it's not so small that resources are limited.

Small agencies have the money to begin implementing a capable team in their organisation but haven't grown so big that the wage role and bureaucracy are unmanageable. You'll still work with a small team, making the whole experience very personal so you get to know the people you're working with. You won't be passed from one professional to the next, which is a risk with larger corporations. Small agencies are keen to invest in who they hire and pay more for top talent than small agencies can afford to, along with putting additional funding into their resources.

Of course, they still won't have the same number of resources as large companies and may still specialise in a niche rather than being a full-service agency that does a bit of everything. If you're looking for a specific range of software to build your product, a large agency might be better equipped to help you out. However, what small agencies may lack in resources, they certainly make up for with their passionate, personal service and dedicated focus.

Using a small agency is slightly more expensive than using a freelancer, as you'll find that decent freelancers are 20–30% cheaper than most small agencies (though some freelancers match their prices to that of small agencies). Either way, you'll need to factor this into your budgeting.

Though this size of agency is likely to have a project management function and processes to make sure the team is executing to the right standard, you'll still need to be involved at key points in the project's development. However, these are likely to be relatively light-touch and focus mainly around making sure you're happy to sign off on each milestone. This is different to some small agencies and freelancers where you're expected to play project manager in the day-to-day interaction between the various functions needed to make your tech idea.

Large agencies

Large agencies are the big dogs of the outsourcing world. They have a much larger team of staff than small agencies, a larger revenue, and a wider range of complementary products and services.

They'll learn what your product's development entails before pairing you with the right staff and may cross-sell services to you from their wide offering. That ensures that what you need can be catered to, and you probably won't have to compromise too much on your product. They'll have a range of top-quality resources and the ability to work on many projects simultaneously, so they can meet your deadlines regardless of how busy they are.

Of course, there are some downsides when working with a large agency. Even if you're allocated an account manager, your personal engagement with those delivering the project can sometimes get lost, and it's common to be passed around from person to person. It's also unlikely that you'll ever speak to the CEO as they'll be focusing on running the business rather than working on tech development. The account manager will hear your requirements and communicate them to the rest of the team. For a lot of businesses, this won't matter as long as the work is high quality, but if you like to engage with everyone working on your project, it could be a deal-breaker.

Nevertheless, you should remember that even if the personal touch is lost a little, a large agency will be more than capable of fulfilling your order and are unlikely to miss deadlines or other commitments you set with them. This is because they're likely to have more protocols and processes to help work run smoothly and plenty of experience working on products just like yours — that's how they were able to grow so large.

If you pick a large agency that focuses on a given niche, they may also have IP in their code libraries that they can re-use or licence to you, meeting your need with something they have already built. However, due to the operational overheads that come with a large agency, their rates are more expensive. Typically, a large agency will be about 40%–110% more expensive than freelancers and about 50%–60% more than smaller agencies.

Closing words

Now you know the different ways you can go about building your tech idea, you can weigh up the pros and cons of each approach to decide which works best for you.

For example, do you value the quality and process advantages of choosing an agency? If you have an internal team, are they readily available, and if not, can you afford to wait until they are before you begin your project? Whereas if you have no money at all, then your only option is to learn to code or use a DIY alternative.

If you're unsure what to do, take some time to research the various options and suppliers. Learn about the culture or people behind each, listen carefully to what they say their values are, and understand their skills, capabilities, and experience. Think about which factors are most important to ensuring the success of your project and consider your limitations.

Key takeaways:

1. Learning to code is the cheapest but most time-consuming way to approach building your tech idea. The same goes for DIY platforms, which can be a cost effective but see you in charge of completing a lot of configuration work.

2. Internal tech teams can be a useful resource for established business if you're lucky enough to have one but may not always be readily available.

3. Partnering with a CTO is a strategic way to build your tech idea but should not be seen as the cheap option.

4. Outsourcing to freelancers or agencies is a very popular way for businesses to approach executing their tech idea but means you need enough money to pay for the development of your idea.

5. The size and nature of the company you outsource to will impact the cost but also has benefits. Pick the outsource company profile that works best for you based on your budget constraints, project management, and business continuity requirements.

CHAPTER 12:
HOW TO PICK A TECH TEAM

"The absolute truth is that if you don't know what you want, you won't get it."
— *Andrew S. Grove, High Output Management*

Chapter Relevance	
New Internal Tool	**New Product**
↑	↑

In the last chapter, we covered the different ways you can build your tech idea, including the various types of suppliers. However, we didn't go into detail about how to choose a supplier within each category. For example, if you know you'd like to work with an agency, the next step is finding the agency that is right for you.

This chapter helps you to decide whether the tech team you are thinking about working with is the right fit for you and your requirements. The lessons in this chapter apply to all options covered in the last chapter, except if you've chosen to code it yourself or use a DIY approach.

I'll cover the difference between submitting a tender that you ask suppliers to bid for compared to engaging suppliers directly, and we'll discuss the pros and cons of both approaches.

I'll then move on to the factors you must consider before picking a supplier, such as price, quality, speed, proposal strength, Intellectual Property (IP) advantage, reputation, and more, giving you a quick list of questions to aid you with your decision.

Finally, we'll close by demonstrating the importance of a well-conducted planning and discovery process and what to expect on this journey.

Technique 1: Tendering

One way to find a supplier is to put your project out for tender, which is a process where you publicly invite companies to bid for your project.

If you work for a government, publicly funded Non-Government Organisation (NGO), or charity, you will normally be expected to prove that you underwent such a competitive process, and you may be scrutinised for the reasons why you chose your supplier.

If you work for a large or public firm, you should check your organisation's policy information to make sure you don't accidentally break the rules.

The tender process

Here is a quick overview of a typical tender process:

1. The request for proposal

The first step of putting your project out for tender is to create a Request for Proposal document, or RFP for short.

This RFP is similar to the scope document we outlined previously but with more sections, such as a background to your business and objectives, the bidding process and timeline for making a decision, the project scope, the format you'd like bids to be submitted in, any critical criteria the suppliers must meet (such as process or size), and the information that must be provided by each supplier.

2. Tender promotion

You will find that there are many different tender portal websites for posting local and national tenders. If you use one of these, your bid will be available for everyone to see. Alternatively, you may choose to opt for a closed-tender process, where you privately reach out to a shortlist of suitable suppliers, sending your request for proposal and asking if they are interested in bidding. If you work for a large company, some will have a list of approved suppliers that you can approach with your proposal.

3. Receive replies

Once you've promoted your RFP, you should expect suppliers to let you know they plan to submit a bid. After some time has passed, you will begin to see them submit those bids to you. Remember, suppliers will follow the process and timelines you set out in your RFP document.

4. Shortlist

Next, you should have a process to shortlist the suppliers you would like to invite to a second stage, which is an opportunity for them to further clarify their offer to you, explaining how they meet your tender criteria.

To be respectful of everyone's time, the amount of work and commitment involved for a supplier to proceed to each stage in the process should increase gradually. Don't expect every supplier to submit a 100-page detailed response if you've not yet confirmed they made your shortlist! Start by asking for a small amount of information to qualify them in or out, then request more information such as a formal proposal later in the process.

5. Decide

You should choose your decision criteria and scoring system, then publish it. It's common for each criteria to have a number of points that can be earned based on the bid characteristics that you value. You should also keep an audit trail of how each supplier scores against your decision criteria.

Suppliers will reply to your tender with your score criteria in mind, so be careful. You don't want to accidentally give more weight to factors you don't care about over those you do as this may force you to pick a supplier because they score the best but is not best suited to meet your requirements.

Example RFP structure

Here's a simple structure you can follow in your RFP:

- Background information about the opportunity and your organisation
- Tender submission instructions
- Technical specification (or your scope)
- Main tender response form (list of questions that suppliers can answer)
- Price requirements (for example, budget range, rates, item prices)
- Terms and conditions of the tendered opportunity
- Award criteria and evaluation criteria
- Appendices

Here is a sample format of the various deadlines and key dates you should define:

- Submission of "Expression of Interest" deadline
- Submission of the formal tender documents deadline
- The timing of any presentations if required
- How long you will take to make your decision
- The award date (when you will pick a supplier)
- Key dates for communication (such as responses to any Q&A sessions)

The pros and cons of tendering

Tendering allows you to demonstrate a competitive process by inviting several suppliers for the opportunity to bid. By offering your project to the whole market, you can find suppliers who have a business approach you may not have thought about and who can meet your requirements extremely effectively.

This competitive process is particularly beneficial if you're targeting suppliers who have pre-built existing solutions on the market. This is because you don't know all of the nuanced capabilities of each out-of-the-box software product. Tendering gives providers a chance to position their offering to the specific needs of your tender RFP information. They can then make a commercial offer to meet your requirements that is not obviously available from their website or standard marketing materials.

However, you may want to think twice before considering this approach for highly bespoke tech projects. The downside of tendering is that it requires a highly structured and formal process, which invites less collaboration or exploration with your chosen supplier.

For example, you need to outline your requirements and decision criteria before accepting supplier bids. If your supplier has a better idea about how to meet your business objectives due to their extensive experience, you may risk discounting them from the process because they don't meet your decision criteria, despite them being the best choice. It may also be impossible to identify fair or appropriate scoring criteria in advance of knowing those better ideas.

For small to medium-sized projects, going to tender can sometimes counterintuitively actually reduce your choice and increase the price. But why?

Think of bidding for tenders from a supplier's perspective. It's very time-consuming and requires the company to have a mature process for finding and bidding for opportunities. Small agencies and freelancers are unlikely to have this process in place, which means they won't bid and you cut them from the process completely and reduce competition.

Another reason is that some requests for proposal can invite bids from 10–20 suppliers or more. The most intense request for proposal documents I've seen required each company to create extensive bids numbering hundreds of pages — that's a big time-commitment for a small business.

Small organisations cannot afford this speculative time. For example, if 20 businesses apply for your tender and each application takes a week of work to prepare, then on average small suppliers would have to spend 20 weeks applying for tenders just to stand the chance of winning a single one. Inevitably, this extra time must be reflected in the price they bid.

Economically, this can create issues for buyers too. Why would anyone spend 20 weeks of work trying to win a very small contact for a few thousand dollars? The cost benefit trade-off doesn't work, and the most cost-effective suppliers might not apply. In practice, it's the larger companies who are most happy to bid for such speculative projects.

Tendering for suppliers for large projects removes some of these cons. If a project contract is worth millions, then the cost of investing 20 person-weeks to create a proposal is worth the cost on average relative to the benefits.

Cards on the table, please be aware that I'm biased against putting your project out to a public tender, and I'm not in favour of a full formal tender process for the supply of highly bespoke projects. [166]

In my opinion, if you want a competitive supplier selection process but also want to mitigate some of these negatives of going out to tender, then I recommend you follow a light-touch approach:

1. **Identify the characteristics that are important:**

   Think about the most important characteristics from a supplier. For example, do you care more about price or quality? Do you want to deal with a freelancer and have one person work exclusively on your project or do you value continuity of supply over the coming years?

2. **Identify a handful of suppliers who meet your criteria:**

   For example, you can do a web search for providers in the area or even internationally. You might ask people in your network for recommendations.

3. **Contact each supplier and qualify against your criteria:**

   Explain your project to them and your initial scope. Understand their process and consider how it aligns with your priorities. If you discover something they do differently that you haven't thought about, you can add that to your decision criteria to compare with other providers. A couple of emails and a 20-minute call should be enough at this stage.

4. **Ask for a ballpark price:**

   By this point, you will have a clear scope for them to review, so they should be able to provide you with a ballpark estimated price. You should expect ongoing discussions, which may change the supplier's understanding of what is required, changing the price. However, getting a ballpark quote means you can judge roughly whether you are on the right page and get a basic number to compare each provider.

   If you have a £30k budget and someone quotes £60k, that might be too large of a difference in expectations to accommodate. Likewise, if you were quoted just £10k for something you expected would usually cost £30k, this might be an indicator of inexperience or elements of the development process that are missing.

5. **Shortlist and refine:**

   Narrow down your shortlist to the final two suppliers based on the calls and emails you've had so far.

6. **30–60 minute follow-up meeting or video call:**

   Arrange a follow-up call with both of the final suppliers to have a longer discussion. An hour is short enough that it respects everyone's time but long enough that you can ask substantial questions and get a feel for whether you can work with the provider.

7. **Commission the planning/discovery process:**

   If your supplier has a dedicated planning and discovery process, then once you've chosen your supplier, you can commission this process with them. Some companies will tie this into the whole project, requiring you to sign off on the full project cost before planning starts, while others will bill for it as a separate piece of work and quote for the project once the planning is complete.

Technique 2: Pick a supplier directly

The other option to build your tech project is to research and engage with suppliers directly. You can find suppliers relatively easily using a search engine, from a recommendation, by visiting trade shows, or on social media. [167]

Later, in "Chapter 15: The 25 Marketing Channels", we will explain the different marketing channels and how to use them to promote your tech idea. These channels are also useful for finding suppliers as those suppliers will use them to advertise their business. So, when you get to this chapter try to reflect on how you could use each channel to both buy and sell.

But what should you look out for in a supplier? And what questions can you ask to decide whether a chosen supplier is a good fit for you?

Here are 14 deciding factors to guide you as you pick a supplier for your project:

Factor 1: Capabilities

Not every supplier is going to be a one-stop shop to provide everything you will ever need. You may have to look around and build your own team of specialists with the capabilities to build your product. Different common capability specialisms in the tech industry include:

Design: Tech designers are often confused with developers, and the terms are commonly used interchangeably ("designer and developer"), but the two are very different. Tech designers don't build the core of your product — they focus on the visual aspect of it and creating something that's entirely user-focused. For example, a user interface (UI) designer will build the front-end visuals, such as the placement of your navigation bar and your colour choices, but they won't build the coding that creates the website's functions. And a user experience designer (UX) will conduct research into how users interact with and use your tech idea. This research carries a cost, meaning not all companies will provide this service but your product may be better for it if you have the money to spare.

Development: Tech developers build the core coding of your product.[168] If you have a vision of how your product will work and what features it will include, your development team will program them into usable elements within your product. A full-stack app developer, for example, will create everything an app needs to function. Your developers will usually work with a UI designer to create and test the visuals, which is the front end of the website the user interacts with. When you hire an agency to develop your tech idea, they'll usually have both designers and developers on board so that you don't have to find two different teams.

Marketing: Once your tech idea is built, you'll need to use marketing to attract your audience's attention. Some suppliers will have an in-house marketing team whilst others will leave this up to you. Marketing and development are very different disciplines, meaning it's normally fairly easy to source a different supplier to execute your marketing. If you look for marketing outside of your development agency, you can either look for a consultancy, a full-service marketing agency, or individual freelancers, such as a social media marketing agent, a photographer, and a copywriter.[169] We'll cover how to market your tech idea in a lot more detail later in "Chapter 14: Marketing Principles".

Full service: A full-service supplier takes your project from an idea to a fully marketed product. They'll do everything you need, and you'll only have to hire one supplier for the entire duration of your product's lifecycle. Full-service agencies are normally large companies with established teams. Some small agencies claim to be full service, but be cautious of this as they may be attempting to do too much relative to the capabilities of their team.[170]

Questions you can ask if their capability stack is important to you:

1. What do you consider your unique selling point is compared to other suppliers?

2. What is the education and experience background of the starting founders? (As this can often influence the culture and capabilities of the company)

3. Do you just do X or do you offer other services to support this too? If so, what?

4. Are your developers in-house and on full-time payroll or outsourced?

5. What different capabilities do you have in your business and who is responsible for the delivery or advice around those capabilities?

Factor 2: Technology stack

The technology stack is the layer of software and tools that the developer will use to develop your product. [171] If you have a specific tech stack in mind — or at least parts of the tech stack — you'll need to find a supplier who can cater to your needs. However, in my opinion, it's best to be open-minded about the tech stack and trust your developer. Instead of being specific on which technologies to use, focus on how the chosen technologies can best meet your requirements. [172]

For example, common tech-stack options for web-based projects include NodeJS (JavaScript), Python, PHP, ASP.net, C#, and Ruby. You don't need to know what all of these languages are but you should know that the decision to pick one over the other can say something about the culture and fit of the organisation you want to work with.

For example, choosing between the long-standing PHP, ASP.net, and C# languages has always created disagreement in the development community, and the use of new languages has caused similar debates in recent years too. ASP.NET is an open-source web development framework developed by Microsoft, and PHP is purely open source with support from other developers.

Both of these languages are widely used by huge enterprises and have brilliant communities, but they are still better suited to different projects. Organisations that follow the Microsoft way are more likely to want to go with ASP.net and organisations that don't may be more likely to choose something different, such as PHP or NodeJS. If your organisation is large enough to have a preference for one suite of tools over another, then you may want to pick a supplier whose tooling and programming languages align with your values.

Unless you're versed in tech jargon, working out which is better for you is tough based on that information, so it's usually best to leave this decision to your tech team. And if you don't have a preference, the technology stack used by each supplier may be of low importance to you compared to other factors.

Questions you can ask if the technology stack is important to you:

1. What back- and front-end programming languages will you use?

2. Are you planning to use a monolithic server structure, microservices, or something in between?

3. Why did you choose this particular combination of technologies?

Factor 3: Market specialism

Some suppliers will accept projects from all industries, aligning their skills and the nature of the problem they solve to their audience. However, other suppliers will specifically work within certain markets and specialise in these areas. You can find specific tech development for the sports, clothing, and pet industries and pretty much anything else you can think of.

Choosing a supplier with a market specialism in your area has perks as it means they've worked on similar projects within your industry that they can draw on for inspiration. If they have target audience research, they can use it on your product without having to start from scratch. If something hasn't worked in your industry in the past, they'll know not to do it again. In other words, they have experience with ideas like yours. You can take advantage of this to speed up your product development and create something geared towards your specific audience.

But market specialism doesn't just have to be in the industry you work in — it can also be based on the nature of the solution. For example, online portals are one solution to many problems, and these problems may be shared by companies in lots of different sectors.

Questions you can ask if market specialism is important to you:

1. Do you specialise in any particular sector?

2. Do you have any examples of past work you can describe or show me that are similar to, or have features transferable to what we're looking to do?

3. Have you learned any big lessons from having done a lot of work in this sector? What are they?

Factor 4: Intellectual Property (IP) advantage

If a company has a lot of pre-existing source code they can use to solve similar problems, that source code is their IP. Intellectual Property Rights (IPR) are therefore the rights that you or your supplier has over the code they've written.

Not all tech companies carry out entire bespoke builds, creating every element and feature from scratch. Instead, some leverage certain features built by others, adjusting them to suit your product. There are lots of advantages to this, including speeding up building, which saves you money as the customer and offers higher-quality elements that have been perfected over time.

The risk of leveraging existing IPR, however, is that IPR are often sold via licenses, which means you don't own the code outright and just have a licence to use the code. Think of it like a legal agreement that says you don't own the code but have the right to use it in your tech idea. Licencing deals are common practice in software, and free-to-use licencing is the foundation of the open-source movement, where people build software for free and make it accessible for anyone to use.

IPR can sometimes be a complex area to navigate. Your development team should make it clear how you can use the product and if there's anything you need to be aware of, such as license renewals or restrictions.

Your development team might have a preference for developing some things from scratch or using existing libraries under a licence agreement. Make sure to ask them what the cost benefit trade-off is of using their existing code libraries compared to coding everything from scratch. Libraries may cost less but you may not own the IP to the licenced code.

Even if you don't own the IPR exclusively, it's possible to have a licence agreement in place that won't limit you. For example, if your tech team assigns you a "perpetual, royalty-free licence to use and extend the code" in a way that doesn't restrict your business operations, then there is little downside — you pay less, they get to bill for some of the code they've already written, and everyone is a winner. [173]

Questions you can ask if IP considerations are important to you:

1. Are you able to reuse any of your existing code libraries to execute this project more efficiently or effectively?

2. Will I own the full copyright to the project code or is some of it licenced to me? If so, what are the licence terms?

3. Do you plan to use any open-source libraries or frameworks to deliver this project?

4. Can I see an example of the IP clauses from your contract?

Factor 5: Price

Even if you don't have a defined budget, the price of a supplier is always important and will inevitably play a crucial part in your decision. Price can tell you a lot about a company, including their experience, reliability, and what you can expect in terms of quality, so don't skimp on research in this area.

Some companies that charge above average rates could be out of touch with the current market. By opting for them without doing enough research, you could end up paying above what you need to for the work they're doing. Of course, they could have a unique selling point (USP) that sets them apart and makes the price more reasonable, such as a high-profile client list or unique software that can give your product an advantage in its market. You have to weigh up whether the benefits are worth the cost.

However, cheaper doesn't mean better and you should be very wary of suppliers that are significantly undercutting their competitors. A cut-throat price could indicate a lack of experience and less commitment to the work they're doing. The ultra-low price might be less because something big is missing. For example, they might put less time into your work and deliver a lower-quality product. Development isn't an area you want to cut corners in, and if someone writes thousands of lines of code the wrong way, they might be creating hundreds of hours of avoidable rework costs for you later in the form of technical debt!

An unusually low price might also be an indicator of an error in quoting or scoping your project. If they charge less because they've missed something significant due to inexperience, you may discover halfway through the project that you have to pay extra for something you thought would be included.

For example, has your development team included line items in their quote breakdown for testing and project management? If the answer is no, then maybe they're cheaper because you aren't getting any testing or project management services. And together these testing and project management services can easily add 25-50% to a project's budget if done properly! [174]

Do you think it makes sense to cut these essential activities out to save costs and risk the success and quality of your tech idea? I don't think so!

Pay attention to price and try to pick a supplier that delivers overall value for money when considering all of these factors.

Questions you can ask if price is important to you:

1. Are you able to give me a detailed itemised cost breakdown?

2. What portion of the project will come from planning, design, development, project management, and testing?

3. Are you able to deliver for a slightly lower price if I can be flexible on the start and end date?

4. Can you approach the technology stack in a different way to lower the cost?

5. Can we avoid a graphic design process to lower the cost if I accept that the interface might not be as polished or easy to use? [175]

6. Does your quote include testing and project management activities? How will you approach these activities?

Factor 6: Key contact relationship

A key differentiating factor that many people overlook is relationships. You'll likely be given a key contact to work with during your project. This might be the freelancer you hire, the CEO of the small agency you choose, or the developer in a large company. Whoever it is, ensuring that you get on well is crucial to your project's success.

You should be able to speak easily with your key contact, sharing ideas openly and comfortably with easy understanding on both sides. Your contact should be professional but friendly and open to your ideas without pushing their own too forcefully. Suggestions are great as they do have the expert experience, but if you don't like their idea, they should respect that and move on.

If you feel comfortable and have a good rapport with your key contact or other members of the team you've liaised with, then the whole project will feel much smoother. [176] You'll waste less time struggling to communicate your ideas and feel happier to discuss your options. Finding someone who understands you and your vision is invaluable when hunting for a supplier.

Questions you can ask if the supplier relationship is important to you:

1. Will we have regular meetings to keep me informed of the progress of the project?

2. Will I be assigned a project manager or an account manager?

3. What can I expect from post-project support once we've launched the first version?

4. Do you believe in my vision for this project? Do you believe it will be a success?

Factor 7: Speed of delivery

If speed to market is your priority, you may choose a supplier that gets to work without delay once the contract is signed, driving your project forward, not slowing it down. However, it may be unrealistic to expect your supplier to be ready the minute you are. If you're hiring a quality supplier, it's common for their schedules to be booked up in the short term and you might have to wait a couple of weeks for work to begin, or even months in some cases!

Rather than worrying about the start date, instead look at how quickly they'll deliver work once they begin. Get an outline of milestones and at what stage the developer expects to be finished to give you a rough idea of the speed of delivery.

It may also be an option to speed up delivery if more of the supplier's team works on your project in parallel, though be careful as the more people who work on a project concurrently, the more time they'll need to coordinate their work, particularly with developers.

For example, you might be able to deliver a project in six months with one developer, or four months with two developers, however, the latter will typically require 10–20% more project hours to account for the additional coordination efforts needed, which will inevitably result in a higher fee to deliver your project.

The project management method that your tech team uses will also have an impact on the delivery speed. Some methods like Agile allow you to start productive design and development work sooner than other methods like Waterfall, which requires all planning activities to be completed in advance of work starting. However, this speed comes at the expense of upfront cost certainty. Make sure you understand how your development team plans to manage the project and make sure their chosen method aligns with your needs.

Questions you can ask if speed is important to you:

1. What aspects of this project can we do in parallel?

2. Can we get multiple developers on this project at the same time if we're willing to pay a little bit more? If so, what is that additional cost likely to be?

3. Can we work using the Agile project management method with a monthly budget rather than delaying the project start date by waiting until planning is complete for the full project?

4. How long do you expect the user-acceptance testing phase of the project to take? (This is the part of the project where you test that the supplier has delivered on the project requirements).

Factor 8: Quality

The thoroughness of a quote is often a big factor in helping you predict the quality of delivery you can expect from a given supplier.

For example, I'd expect to see a significant portion of your project cost (often 25-50%) assigned to testing and project management. As mentioned previously, if you receive quotes from suppliers that are cheaper and faster but don't include a provision for project management and quality assurance, this may be a red flag that they're cutting corners to present the quote you want to hear rather than the quote you need to hear.

The last thing you want is to get 80% through the project to find that this omitted time and cost is required but wasn't planned for.

It's also the time to think about the size of the supplier. As I said in the last chapter, a medium or larger agency is likely to have better processes in place and more resources, so they're better equipped to ensure quality vs. a micro agency or freelancer. That doesn't mean you won't get a high-quality product from the latter options, but it's certainly more of a risk. So if quality is important to you, then make sure you ask questions to get an idea of what processes exist to ensure a high standard of delivery.

The processes that a supplier uses in the onboarding process are another sign of quality. If a supplier has a thorough planning process with clear documentation, commits to action dates, and keeps you informed, chances are that this standard will be maintained throughout the delivery phase too.

In contrast, if a supplier gives you a very thorough plan and designs for a highly bespoke project before they've even spoken to you, this may make you question their approach. How can a supplier guess what features you need for a highly bespoke tech project after a short call? They can't. And for them to try means they're making price estimates based on rough guesses, which likely means they have poor processes for planning your project. There's a high risk they won't be thorough in the delivery and might build and release features that are what they thought you wanted but don't really deliver on your expectations.

Questions you can ask if quality is important to you:

1. What processes do you have to make sure the project meets my expectations and is high quality?
2. Do you use any version control tools? If so, which?
3. How will you approach testing?
4. What portion of the project is assigned to project management and testing?
5. Once the project starts, what are the day-to-day project tasks that happen behind the scenes to deliver the project plan?

Factor 9: Business continuity

What would you do if your tech provider were hit by a meteor and everyone in their team vapourised in seconds? Could your business continue to operate or would your tech idea be in major trouble?

This is business continuity. In the worst-case scenario where everything goes wrong, what do we do and are we able to recover?

The bigger the company, the more likely they can ensure their operations aren't affected by unplanned incidents. They'll have protocols in place for downtime, staff to cover operations, and money to replace resources that fail.

Most medium and large organisations will have business continuity plans in place that outline precise measures they'll take in cases of emergency. For example, if you know that two or three developers in the organisation have worked on your project, there's a good chance that the company can continue to support you if one of them left.

Very small businesses or freelancers are, unfortunately, more likely to be affected by downtime or unforeseen circumstances that could impact on the continuity of your project. They won't have the same backup resources in place as their budget is tighter, and they'll be more at risk of employee downtime in a smaller team. If too many staff become ill at the same time, projects are likely to become delayed. It's also a lot more likely that a one-man-band freelancer will have multiple projects running in parallel, whereas larger organisations are more likely to schedule dedicated resources to your project.

And if you decide to use an offshore agency, business continuity is at higher risk. Some countries are simply more prone to natural disasters, like earthquakes, flooding, and internet outages, which may increase downtime. If you opt for cheaper labour, remember that the company may have less of a budget and take shortcuts to save money, and a business continuity plan could be low down on their list of priorities. Of course, not every offshore agency carries these risks, but it's best to be thorough with your checks if you do decide to go offshore.

Business continuity can be protected with processes too, for example, where your data and servers are hosted, the backup and recovery processes, or the version control approach (saving historic copies of source code).

Questions you can ask if business continuity is important to you:

1. How large is your design/development/project management team?

2. How will you split resource scheduling across your team?

3. Will I have a dedicated developer or project manager assigned to my project?

4. Does everyone in your team have the skills to work on our project, or are there complexities that rely heavily on the knowledge and experience of one person?

5. Once the project is completed, can we be given access to the source code and data?

6. Will the code be commented so it's easy to read?

7. Can you write a quick-start readme to help with handover should you go out of business?

8. How long has your business been trading for?

9. Is there any budget in your quote to create technical design documentation? If not, then how much would it cost to do this?

10. Will you develop using commonly accepted standards and frameworks?

Factor 10: Process strength

Every company has different processes. Even if the end goal of what they aim to create is the same, the way they get there won't be the same.

Solid processes are critical in business, especially for complex projects. Processes are how a business ensures it can deliver a consistently high-quality customer experience. I know first-hand as I've grown my business that whenever you experience a problem, you solve it by implementing a better process. This means that over time, you build up lots of unique ways to do things, each with its own how-to documentation. For example, we have processes for how to speak to customers, how to deliver a great experience, how to test a project, how to hand over a completed project, plus quoting, quality assurance, and the list goes on! If we didn't have these processes, things would quickly become a mess.

The different processes that your supplier has (or doesn't have) will affect many things, such as the speed they finish your project, the quality of the finish, and how they work with you. It's also vital that the supplier you choose follows industry standards throughout their processes. If you're suspicious that they won't follow best practices, you may want to reconsider your selection.

Look out for signs that your chosen supplier has good processes. Maybe they send you slides explaining the steps to work with them or ask you to complete a short form to kick off the project with them. Things like this demonstrate that your chosen supplier has standardised ways of working they've optimised over time working on many different projects. And if you don't see processes in places where you think you should, then take note, as there may be processes missing elsewhere in the delivery process too.

Questions you can ask if process strength is important to you:

1. Have you documented your project and development processes?

2. Will you conduct a planning and discovery phase?

3. What happens once development is completed?

4. Do you follow any best practice approaches in your industry? If so, which?

5. What project steps or milestones can I expect?

Factor 11: Proposal strength

Whether it's before or after planning, suppliers will often present some form of proposal outlining the project goals, their deliverables, cost, and deadlines. Sometimes these proposals are highly detailed and accurate and sometimes they're a rough outline of the general direction of travel to set expectations and get your buy-in.

A good proposal should be detailed with useful information (no fluff and waffle!) and let you know exactly what you're going to be paying for. Once you've narrowed your suppliers down to a few different choices, reviewing their proposals is a great way to help you set them apart and pick the right fit.

Questions you can ask if proposal strength is important to you:

1. How accurate is the price and/or timeframe in this proposal?

2. How did you go about deciding on the approach outlined in this proposal?

Factor 12: Openness and transparency

How transparent is your key contact in their proposal and throughout your communication so far? For example, are they open about how they're going to approach the project, how they'll fulfil your needs, and how they break down their costs?

It's a sign of trustworthiness when a supplier is open with you about their processes. Look out for things like cost breakdowns and detailed step-by-step processes to reach your goals. All of these snippets of information come together to build more open communication and trust between you and the supplier.

Questions you can ask if openness and transparency are important to you:

1. Are you able to provide a detailed, itemised quote?

2. What happens if we realise midway through the project that something I wanted is missing from the brief?

3. Are you confident that you have the capabilities to deliver these project requirements?

4. Will I be charged for bug fixes that I raise when we get to the User Acceptance Testing (UAT) phase of the project?

Factor 13: Pricing model

I've mentioned pricing a few times, but now it's time to look at the pricing model of your chosen supplier. There are several ways the cost of your project can be presented. These include:

Blended rates: If you're working with more than one member of staff and they have different hourly rates, you might be asked to pay a blended rate. This is when the rates of all staff are averaged out to give you one sum, making it simpler for all involved. Blended rates can also apply when a job entails many different styles of work that the individual would usually charge different rates for. Rather than provide a breakdown of an hourly rate for each, they'll create a blended rate from an average of them combined.

Tiered rates: Tiered pricing is where different functions in the agency are billed at a different hourly rate. Tiered rates are less important for a highly specialised agency that focuses on one area of service delivery but may be more important for agencies that provide a range of diverse services as you wouldn't expect to pay the same for certain marketing services and development services.

Intellectual Property charges: If you're buying software from a third party or if your supplier is using their code libraries for your project, there are two options: licensing and ownership. If you're licensing software, somebody already owns the intellectual property rights (IPR) and can dictate how you use the product. If your developer is using features and capabilities that they or a third party has developed, you may have to pay the license fee, otherwise they may have to quote you more to build them again from scratch.

Package pricing: If you receive a single price figure from your supplier, this is package pricing. All work involved in developing the tech product is grouped into one lump sum. Sometimes, suppliers will offer a package price that's cheaper than each individual piece of work, encouraging you to give more of your product development to them. But package pricing can also be risky if you're not sure what you're paying for. If you suspect that you're paying too much, ask for an itemised breakdown of the lump sum.

Itemised: An itemised pricing model shows how the supplier has priced up the project: task by task or requirement by requirement. For software projects, the list can include the individual prices for planning and discovery, idea meetings with you, software development, feedback meetings, and a round of edits based on your feedback and customer feedback. You get to see exactly what you're paying for and understand the value in the total figure, making it a popular pricing model method.

Questions you can ask if the pricing model is important to you:

1. What is your process for coming up with a quote? How do I know it's accurate?
2. Are there any licence costs included in your quote?
3. How will I be billed for post-project support?
4. How will I be charged for adding additional capabilities to the project after the first version is completed?

Factor 14: Reputation in the market

A supplier's reputation in the market is a form of social proof that demonstrates they can deliver on expectations, where the proof doesn't just come from what *they* say they can do but from what *others* say they can do.

You should consider whether the supplier has an established brand, and whether people's perception of that brand aligns with what the company says people think about their brand. You can learn about a company's reputation by looking for reviews online or asking for examples of their previous work where they can provide a statement of customer feedback. Look for signals that the marketing materials you've read genuinely represent the company's capabilities and values.

Once you've evaluated the potential supplier(s) against the criteria, you might find that there are some trade-offs. For example, maybe they're more expensive but have the track record to prove they can deliver quickly or to a very high quality. You'll need to weigh up the pros and cons and decide based on which qualities matter most to you.

There can be significant trade-offs. For example, there is a famous saying in the project management world: *Time, cost and quality, pick two:*

Be mindful that some of the things you might look for in a supplier mean that other things aren't possible. For example, if you want the highest quality end-product possible, this will probably mean it won't be the cheapest solution available.

Questions you can ask if supplier reputation is important to you:

1. Do you have an example of similar past projects you've worked on?

2. Do you have any written client testimonials or reviews that you can share or link to?

Advice if you must source several quotes

If you work for a large company or public department, you might be required to prove that you got several quotes from suppliers and followed a competitive process. A request for three comparable quotes is typical in this scenario.

Getting several quotes can be a major headache involving lots of time and energy coordinating many suppliers, and you likely won't have the time and budget to proceed with a planning process for all three to get accurate quotes, so what should you do?

If this multi-quote scenario applies to you, then you may want some way to compare each supplier on price and approach without wasting too much time on unnecessary processes. In this scenario, begin by asking for a ballpark price based on your scope and make it clear that you appreciate this price could change once more information is known. Another approach is to ask the supplier how much projects similar to yours cost as they should be able to answer that question more easily.

The fastest way to ask for three quotes is to already have a rough idea of the amount time you'd like a project to take, such as if you know your project must not take longer than six months to plan, build, and launch.

In this scenario, your launch timeframe will have implications on how much work you can realistically hope to complete in that window of time. You could then ask a question like "How many billable hours/days would a 6-month project normally take?" Then, based on the answer, you could ask how much that amount of billable time would nearly cost.

Then, if you want to get a second or third quote quickly to make up your quota of three, you could ask other suppliers for a cost to deliver a given number of billable hours or days, which is a reply they will be able to provide very quickly compared to if you asked them to audit your full project requirements.

Planning and discovery

Earlier in "Chapter 10: Document Before You Implement", I discussed the scope, specification, and wireframe documents you can create to outline your project. The planning and discovery process further develops the early ideas you wrote down when creating your project scope.

I appreciate that it may feel like the act of putting together your scope and wireframes is "planning". This is true — they are a form of planning. However, when I talk about planning and discovery, I mean the process where your team and chosen supplier learn and share ideas with the objective of identifying a detailed plan of action to build your idea.

If you've taken the time to write your own specification, then this will certainly help the process, but don't be surprised if your chosen supplier also wants to create their own documents to support their planning process. Don't worry — it's a good thing for them to also want to create their own documentation and your ideas can be approved in the planning process with them.

Your chosen tech team will be familiar with scoping projects and will consider things you won't have. Your supplier will create high-quality planning documents that remove ambiguity about your requirements and increase the chance that your project is successful.

Every agency approaches planning and discovery differently, with some opting to do digital whiteboarding sessions and others who have a series of short calls or meetings to complete the specification in parts.

Personally, I prefer to keep the first planning meeting as personal as possible. I start with a meeting in person or over a video call, and have members of our design team on hand to provide visual input and our project team to take notes so I can focus on leading the discussion.

This personal approach removes barriers from the conversation and allows everyone involved to build a social relationship with each other. Often, the relationships built in planning will shape the effectiveness of the future collaboration, so it's just as important that everyone leaves the first meeting feeling that they like one another as it is that the project is well-defined.

I've found that the most effective planning sessions include as few people as are necessary, normally two stakeholders on your side and two or three from the supplier.

The people involved in planning should be those who have a vested interest in the project, power within your company to make decisions, and enough domain knowledge to answer any questions. This doesn't mean that others can't be involved, but you may want to include them in a separate shorter follow-up meeting to get their input on the bits that impact them the most, things they have the most expertise to assist with. It's important to respect everyone's time.

The number of planning meetings needed will depend on the size and scope of your project, with longer more technical projects requiring more time than shorter, simpler ones.

Here is a typical planning agenda that can be completed over one or more meetings. These meetings can happen in person or over a video call.

1. Introductions:

The most important goal of the introduction is for everyone to meet, say hello, and get relaxed in each other's company. This will usually include small talk and making drinks. Once everyone is relaxed, you should move on to discuss each person's role in the business, yours and your supplier's, and how your role relates to the project.

2. Business background and overview of the problem to be solved:

You work in your business day in, day out, so you have working knowledge about the way you do things that your supplier doesn't, and it's important to bridge that gap. Even if you're working with an internal tech team, their daily activities will look very different to yours, so it's important that they empathise with what you do.

Use this section of the meeting to make it clear what the business does, or plans to do, and what opportunities resulted in you wanting to take on this project. Explain how your role fits into this vision.

3. User journey analysis:

Your tech team should start asking questions about how you currently do things, reflecting on your new project scope and where it fits into your business objectives. This will happen in loops, going end to end through different parts of the plan, then reflecting on how each part slots together.

They should aim to see the world from your perspective so they can apply their technical knowledge to challenge your scope if required and propose solutions and features. This is where their technical expertise should shine and you'll be made to think about eventualities that you hadn't considered.

If you have customer feedback about your idea or have conducted any pre-emptive user interviews, then now is the time to share your findings with your supplier.

If your system integrates with other systems and data sources, then you'll discuss the different data journeys and user journeys. This is the process of identifying the varied ways that users and stakeholders will navigate through your end solution.

It's important to understand these flows and journeys because each step in the process represents a new requirement for your system. The "what if" questions that come up when you consider this process will highlight requirements you may not have thought of.

4. Live wireframing / visualisation:

Your supplier will want to sketch what they have in mind. These sketches may include user interface wireframes, flow charts, database and data structures, critical paths, sequence diagrams, or other software design diagrams such as UML diagrams. [177]

Any wireframes or diagrams created in planning should be strictly in sketch format, done on an in-person or digital whiteboard, or on paper in front of you. Keep it light-touch, enough to communicate ideas rapidly and not slowing down the flow of the meeting. They can always be improved upon afterwards.

If planning is conducted over several meetings, then your supplier may want to create wireframes or other plans in between meetings. This is an acceptable alternative approach.

5. Re-run of the process:

This stage of the process is to ensure that all stakeholders have a shared understanding of what is required.

Once you've gone through each section of the scope and discussed what you already have, the problem you're aiming to solve, and the possible solutions, it's time to provide an overview of everything discussed. Quickly run through each user journey and how it fits into the project, then reflect on whether there is anything missing or not discussed.

6. Plan the next meeting and actions:

Depending on the size of your project, one planning session may not be enough for your tech team to gather enough information to be able to complete the full specification and wireframes.

I find that the best meeting length is highly subjective on who is involved in the meeting. Some people can sustain a high level of concentration for much longer than others.

The number and length of meetings required to complete planning will significantly impact the lead time of the planning process. So, there is a fine balance between meeting length, planning timescales, and planning effectiveness. If you can conduct planning in fewer meetings, then you probably should — providing everyone involved can maintain a healthy level of focus and concentration.

When you inevitably hit the three-hour mark or concentration starts to slip, call the meeting to an end. Your tech team should reflect on where you've got to and what items still need further discussion. These outstanding topics will form the agenda for your next scheduled meeting.

7. Follow-up meetings:

For medium or larger projects, you may require several meetings, where the wireframes, diagrams, and specification are released to you in phases to sign off.[178] For example, I typically ask our clients to sign off on any wireframes we produce and have a handover meeting to show what we've done before we start writing the requirements specification.

8. Proposal:

Once you've completed the planning process — and the specification and supporting documents are signed off — your tech team should have what they need to let you know the resources required to deliver your project. That is, how many development/project hours, how long the project will take, and what it will cost.

Suppliers have different ways of presenting their proposal and cost breakdown, so expect to see it in large sections or itemised by feature. I recommend that you request a granular breakdown, as this gives you confidence that they haven't just made up a price from thin air and you can see the cost of each individual requirement.

I've had clients ask for 'small' requirements that turned out to be a few full weeks of development work, and by showing them the itemised quote, they could make a commercial decision about whether the requirement was worth it or not. Itemised quotes allow you to decide what your MVP should be by considering the impact of each feature relative to the effort (its cost).

However, be aware that the project management method chosen will impact on your supplier's ability to create an accurate itemised quote. For example, if your supplier recommends an Agile approach, then their 'quote' at this stage is likely to be a rough ballpark price that will shift and change as the project progresses. We will cover more about the differences between Agile and Waterfall project methods in the next chapter.

Some organisations choose to avoid planning and discovery due to time pressures to deliver, no management buy-in, lack of experience in conducting a discovery phase, or lack of resources or funds to conduct the activities required. To understand why this is a bad move, let me explain why planning and discovery is so important.

Discovery is a process of learning where you reflect on the business environment to make better business decisions. Planning is a decision process where you clearly and concisely document what you decide on.

This process is important because you're best enabled to plan how to build and execute your app if you first take the time to understand the environment which you'll operate in. You must understand the business needs, speak to stakeholders, and challenge existing ideas or the status quo to build your perspective on this business environment. Once you've finished this discovery-led learning process, you'll be ready to set out the actionable steps required to achieve your goals.

Research shows that conducting discovery can reduce your risk of failure by 75% or increase your chance of success by 59%. [179] Do you think that the time and cost of planning is worth it to achieve a 59% increased chance of success? That sounds like a great trade-off to me. Given this incredible ROI on planning and discovery, I'd highly recommend that you don't cut this process out to save on costs.

Once the planning process is complete, your chosen tech team should complete their versions of the specification documents. Depending on the level of service you've chosen from them, it may include wireframes, designs, or other technical diagrams. [180] At this stage, it should be possible for your tech team to provide a very clear plan of action, however how they bill you for their time or schedule your work will depend on their chosen project management method.

User interface and user experience design

If your tech project requires some form of user interface (UI), then your supplier will likely include provisions for a graphic design process in their quote. Sometimes, providers will increase what they charge for planning to do some design work before the main development project, while others deliver any graphic designs as part of the main project agreement. There isn't necessarily a right or wrong way to do this, and I've seen both approaches done successfully.

You will also see the terms UI and UX thrown around interchangeably.

User interface (UI) design usually refers to the process of completing the user interface designs, such as creating visually appealing mock-ups of how each screen should look.

User experience (UX) is a much broader category of service focused on trying to optimise a tech solution to improve the user's experience. UX can include activities such as interviewing users about their feedback on the wireframes or UI design, watching how people interact with the software or software prototypes when unprompted, considering market research or best practice to influence the structure or approach, or writing well-considered copy that makes it clear and easy for users to interact with your tech. You'll notice that we covered some of these techniques as tools to validate your MVP and ideas.

Everyone in your business should have some responsibility for UX, not just the designer! You should try to nurture a culture where everyone strives to optimise and improve the customer or end user's experience.

If you'd like to try your hand at incorporating some basic UX approaches into your project, then when you receive wireframes or graphics from your tech partner, don't just review them yourself — get a prospective end user involved and see what they think about the plans, then use their feedback to adjust your approach.

If your user interfaces have considered the user experience well, end users may be able to figure out how to use your tech with no training or guidance. UX is critical to your success because the more easily users can interact with your tech, the more likely they are to use it and value it, and the more likely you are to meet your wider objectives.

Closing words

You should now have the tools to consider how best to source your supplier and whether you should put the work out to tender or reach out to individual companies who meet your needs. [181]

Once you've chosen a supplier who feels like the right fit for you and your project, review each of the 14 factors and decide which you value the most. Then make a list of the questions to ask your supplier and see how they perform against each factor.

Key takeaways:

1. Tendering a project and picking a supplier directly are very different approaches, and you should pick an approach that best suits your project size and objectives.

2. There are many different types of suppliers ranging from freelancers to micro or large agencies. Each type has a different cost-benefit trade-off.

3. Planning is critical in ensuring the best chances of success for your project and can reduce the risk of failure by 75% or increase your chance of success by 59%.

4. Though you should create your own planning documents, such as a scope or specification, your chosen tech team may still want to create their own too. This is normal and shouldn't be considered a waste of time as it builds and improves on your initial ideas.

5. Consider your delivery time, project cost, and quality priorities — pick two as it's unlikely you can optimise for all three metrics.

6. Not all agencies have strong software development capabilities, so ask questions to ensure that your chosen partner can deliver your project to a high technical standard.

7. Agencies come in varied shapes and sizes, with different cultures and specialisms. If you plan on outsourcing to an agency, make sure you consider the pros and cons of different-sized agencies relative to the various hourly rates billed for design and development work.

8. UI and UX are similar but subtly different disciplines.

9. Use the questions alongside each of the 14 factors to decide whether a supplier is the right fit for you.

CHAPTER 13:
PROJECT LIFECYCLE TYPES

"A traditional project manager focuses on following the plan with minimal changes, whereas an agile leader focuses on adapting successfully to inevitable changes."
— Jim Highsmith,
Agile Project Management: Creating Innovative Products

Chapter Relevance	
New Internal Tool	**New Product**
↑	↑

In this chapter, we'll discuss the different approaches your tech team can take to manage your project, where each option depends on the size, nature, or the preferences of your technical team.

I've hinted throughout previous chapters that there is a big difference between Waterfall and Agile methods of project management. Now it's time to go into more detail about what these approaches are, when they work, and when they don't.

You need to decide which style of project management is right for you before you start. Choose wisely, as you risk losing control of costs and deadlines if you make the wrong choice. The project lifecycle type that is likely to work best for you will closely align with your current ways of working and business culture. You will need decide which you value most: the ability to pivot and change priorities easily or having certainty about costs upfront.

The Waterfall method

Waterfall is a popular project management approach that involves completing each stage of your project in a specific order. You plan everything at the beginning before completing all of the work without going back and changing those plans.

To help you visualise this, this is what a typical waterfall project management process looks like, with each rectangle representing a stage in the project process:

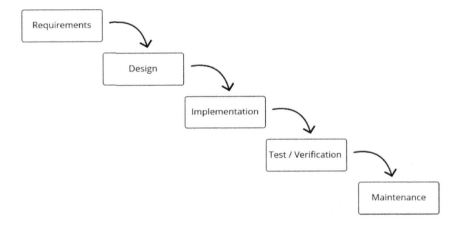

The idea with Waterfall is that you work through each step of the project in isolation, complete it, and sign off on it before moving to the next step. Once a step is complete, you do not re-visit it.

Here is an overview of each of these steps.

Step 1: Requirements:

You start by defining the project requirements, which might be your initial scope or the functional or non-functional specification documents. I also like to include the project wireframes as part of the requirements specification, although some may consider this to be part of the design phase.

Step 2: Design:

This where you create assets that demonstrate how the project should be built. It can include graphic mock-ups, function specifications, flow charts, and other diagrams.

Step 3: Implementation:

This is where your tech team do most of the development work, implementing the requirements and design item by item. I recommend that you have regular check-in meetings throughout implementation so you can be updated on progress and check that the requirements are on track.

Step 4: Test/verify:

Once the implementation is complete, you move on to the testing phase. This phase aims to check both the quality of the solution and that it meets the agreed requirements and design.

Testing may also have several steps, and your tech team should complete a round of testing before they hand it over to you to test. The part of testing that you will be involved in is called User Acceptance Testing (UAT for short) and it's important to give you the opportunity to check that the requirements outlined at the beginning are met.

Sometimes, testing is completed entirely at the end of a project, and sometimes there are smaller rounds of testing at each project milestone. Your tech partner will be able to advise on which approach they feel is best, as it can often depend on the nature of your project.

For example, if a project has many interfaces and sections, then it's easy to complete each section in sequence and release it to you for testing once each section is ready. However, for smaller projects where each feature is closely integrated with every other feature, it may not be possible for you to test until the full project is complete.

As you test your project, your tech team should give you a way to report issues, such as a missed requirement or a bug or fault.

Once you've completed UAT, you will be asked to sign the project off as complete, meaning your tech team have delivered the project to meet the agreed requirements, and it can then be released.

Step 5: Maintenance:

The project activities to support your tech project don't stop once you release it, and you should expect to allocate some time each month to ongoing support and maintenance.

For example, if your tech idea is a mobile app, it will need updates to meet the latest app store rules; the server software needs updates to receive the latest security patches; desktop software needs a refresh to continue working on the new operating system; and any backups of your data need to be hosted and maintained.

Most projects need a post-release support arrangement, even if it's only a few hours each month for your tech team to be on standby. Or if you have an internal tech team, make sure they're aware of the level of support required for your project so they schedule the resources needed ahead of time.

If you're outsourcing the build to a third-party tech team, then your Waterfall cycle will include a few more steps for quoting and contracting:

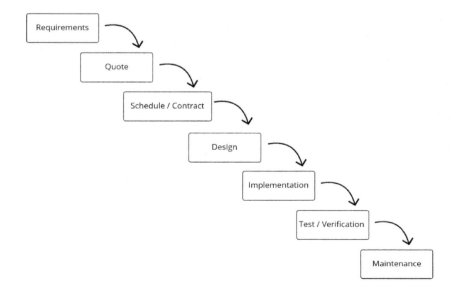

In a Waterfall project, quotes are typically provided before or after the design phase. Quotes consist of the itemised chunks of work needed to deliver the requirements, and providers are normally prepared to lock in a fixed price as long as you don't change those requirements later.

Where quotes are based on time, the project schedule is influenced by the quote as the quote determines designer and developer time to deliver the requirements. Naturally, a 2,000-hour project will take more time and resources than a 200-hour one.

Your tech team should consider every element of your project delivery needs to prepare an accurate schedule. This will include the time spent by their account managers, project managers, designers, developers, testers, and any other role required.

Why Waterfall?

Waterfall is a simple and effective approach to manage small projects that is easy to understand and communicate to everyone involved in the project.

The biggest benefit of the Waterfall approach is cost certainty at the beginning of the project. As Waterfall requires you to plan everything upfront before you start, developers can create detailed quotes that consider every small detail of those plans.

However, this benefit comes with a major downside: flexibility to change. The Waterfall method is not appropriate if your priorities and requirements change after development begins.

The worst thing that can happen with Waterfall projects is when a long phase of delivery is started, only to need a change in requirements midway through. If this happens, then you must undergo a complex accounting process to figure out what's new, what's changed slightly, how much of it is rework, and what the implications are on the quote to deliver the work. I've worked on projects where we've had to do this and seen how you can end up with an almighty mess if you're not careful.

You can mitigate many of the downsides of Waterfall by keeping your requirements small and your MVP minimal. The shorter your project, the less likely it is to change midway through.

Many of the small businesses I've worked with like to use the Waterfall method for the first MVP phase of their project. This is because the requirement for cost certainty usually outweighs the benefits of scope flexibility. Then they switch to Agile for ongoing project requirements after the first release. By definition, their MVP is their smallest feasible set of requirements, so if there are changes, then those changes can usually be treated as bolt-ons at the end rather than causing substantial rework.

If you plan to complete a small amount of work over a short time, then Waterfall is for you. If you need a consistent amount of work every month for a long time, then the Agile method is what you need.

The Agile method

The Agile project management method is a different and iterative approach to project management that sacrifices upfront cost certainty of the whole project to achieve flexibility and speed to schedule resources. The Agile approach is useful for projects where flexibility to adapt and change is more important than cost certainty of the whole projects.

These Agile iterations are called Sprints, and each sprint can be anywhere from one to four weeks but rarely more than that. You plan what you do before the sprint, deliver the sprint, then test and release at the end — before repeating this process for the next sprint.

Here is a visual representation of working in Agile sprints:

The Agile Method

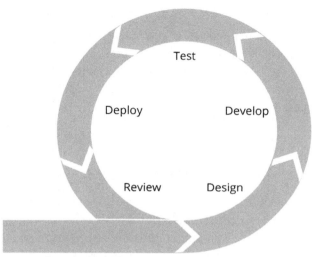

With Agile, you have a high-level vision of where you want the product to be over the coming months but don't define the fine detail of each cycle until the start of the cycle. The process is structured so that you are allowed to change your mind at the start of each Sprint. You only ever lock yourself to deliver the requirements planned for the current sprint and can adjust your plans for what will go into the next sprint before it begins. Your forecast of what features you predict you'll develop in future sprints is called your Product Roadmap. We will go into more detail about how to plan your roadmap in the upcoming "Chapter 17: Plan Your Product Roadmap".

By this point, you should understand why it's important to involve your customer or target user in validating your ideas. The sooner you can do this, the better. However, if you show prospective users your in-development system, you will learn about their values and expectations. You will discover features you didn't realise you need, and others that you thought were essential may not be as important as you originally imagined. Agile gives you the power to adjust to this feedback in ways that Waterfall can't.

Here is an example where a client chose the Waterfall approach to achieve cost certainty when the Agile approach would have been a much better option:

A few years ago, we worked with a client who we'll call Big Corp Ltd. Big Corp wanted to develop a large software product that analysed lots of data, creating dashboards to enable their staff work more effectively. They had big visions of what they wanted their tech idea to achieve and had interested investors.

Big Corp secured the first round of investment and engaged us to build the project, which given large scale of their requirements, we estimated would take approximately 12 months.

Despite the project being very large, the client wouldn't budge in wanting to plan the whole project upfront as per the Waterfall approach. They wanted us to provide a fixed price to deliver the requirements — high-level estimates weren't acceptable. I explained the differences between Waterfall and Agile, and our client assured me that Waterfall was suitable as there would be no changes as the project progressed.

The size of the project meant a lot of planning — and while we normally like planning to take no more than four weeks, this one needed more than two-and-a-half months! The supporting technical specification and other project documentation were intensive to complete. On reflection, knowing the size of the project, I should have refused to work in a Waterfall way, but I didn't — we live and we learn.

We got started on the project and sure enough, at the three-month mark, the investors wanted to see our progress. The client gave them access to our staging demo environment so they could see what we'd done so far.

Having seen the value we had built in the demo environment, the client realised they could give real customers access right away — nine months ahead of the original schedule! They asked us to pause development on new features to complete testing on the three-month version so they could release it to real customers, which we did.

Once real users had access and the real-world feedback started rolling in, the client's priorities went out of the window, and the nine months' worth of development that was planned, documented, and still underway was no longer a priority. Uncoupling the new plans from the old ones quickly became a mess.

Remember, we had agreed on a fixed price quote to deliver the project requirements, so you can imagine that we now had to go through an accounting process to work out what we had built partially and in full, which bits of the specification each related to, and how they were quoted.

We removed features that the client decided they no longer wanted, inserted the new features, and updated the costs and budgets as we went. The process wasn't simple as we'd already started working on some of the features.

In the end, six months of planned work was shelved or replaced through several rounds of adjustments and changes, which was not only a waste of planning effort but also a very inefficient and ineffective way to manage the project.

Agile flexibility removes waste. As you only lock-in plans one sprint ahead, you don't risk throwing away months of specification writing time every time you adjust your requirements.

The quality of your tech idea can also be higher when working in an Agile way because the testing process is also completed sprint by sprint. This iterative approach means that bugs are identified not long after the code is written. This is useful because fixing a bug close to when it was developed means it is still fresh in the developers mind, it's easy for them to pick up where they left off to resolve any issues. It's much harder to debug a problem if you're coming back to something that was built six months ago compared to just a few weeks ago.

Agile also has a different approach to recording requirement. Rather than working from a technical specification document as outlined in "Chapter 10: Document Before You Implement", your tech team will create many micro work packages called tickets. Each ticket is a self-contained chunk of requirements and related information.

These tickets are recorded using specialist project management software. [182] Tickets of work can represent bugs, features, improvements, tasks, or groups of work called Epics or Stories. I will also expand on these concepts in "Chapter 17: Plan Your Product Roadmap".

Sprints are also billed differently to fixed-price waterfall projects. Instead of agreeing on a fixed price to build the fixed project requirements, you agree on a fixed price per sprint and request a certain number of resources from your tech team in that sprint, allowing requirements to change. It's good practice to roughly estimate the number of sprints required in your Roadmap to meet a your vision for your tech idea. [183]

Closing words

If your tech idea is a brand new project, I recommend that you try to keep your requirements small enough to follow the Waterfall method to deliver your first MVP.

Ideally, the development phase of the first version shouldn't take longer than three to six months of development work. Fitting this tight window of time may require you to reduce what you'd like to build so you don't have to plan out lots of phases of development until after your first version is ready.

If you can't find a way to cut your MVP to six months or less, then it consider dropping Waterfall completely and begin immediately with Agile instead. [184]

If you do opt to do Waterfall first, once the first version of your project is complete, it's usually best to switch to an Agile way of working. However, if you don't need a constant flow of updates and changes then you can also treat subsequent rounds of changes to your tech idea as mini Waterfall projects. Though you should only continue to work in Waterfall if you require cost certainty for each release, are happy for planning to take longer, and are certain that your requirements won't change mid-project.

Key takeaways:

1. Waterfall projects require you to define all of the project requirements upfront and deliver those requirements, making little to no changes as you go.

2. Waterfall is useful for projects with tight cost constraints but might take longer in planning before you can begin work.

3. It's more difficult to assign costs to long-term requirements with Agile, as although you might know the implementation time for each sprint, you don't plan in detail much more than one sprint ahead.

4. Agile is a useful alternative project management approach that allows for iterative change as the project progresses.

5. Whether Agile or Waterfall is right for you depends on your company culture, size of project, and stage of project.

6. Consider working Waterfall to deliver your Minimum Viable product, and move to Agile after your first release.

7. Your product roadmap is your long-term vision of what you would like to see in your tech idea over several months or sprints of development. This roadmap is also allowed to change with each sprint.

PART 3:
LAUNCH AND MAINTAIN
YOUR IDEA

Releasing your idea into the wild, marketing
to gain users or customers, and managing
your ongoing feature roadmap

CHAPTER 14:
MARKETING PRINCIPLES

"The more you tell, the more you sell."
— David Ogilvy

Chapter Relevance	
New Internal Tool	**New Product**
→	↑

About the bonus marketing chapters

Execution isn't just about planning and building the project. If your project is to create a tech product, then the way you choose to launch that product can be as — if not more — important than the capabilities of what you've created.

Although I have direct experience of running digital marketing campaigns, my experience and track record is limited to my own businesses. Fortunately, I'm lucky enough to have a father, Steve, with nearly three decades of experience working in marketing and senior decision-making roles for medium and large organisations. Since 2017, he has focused his efforts on helping small- and medium-sized enterprises to achieve their commercial goals through strategic marketing and mentoring, with a focus on technology companies.[185]

Steve kindly offered to contribute these bonus chapters outlining the various ways that businesses like yours have achieved marketing success, driving demand and sales for their tech solutions.

Even if your tech idea involves creating an internal tool rather than a tech product or service, I still recommend that you read this chapter. This marketing knowledge will help you better understand the ways your company can reach its target customers, which might also help you think creatively about how you can innovate and create new ideas in this element of the business too.

Steve's marketing chapter, we'll continue with the final chapters where I'll explain what's involved in managing a development roadmap for your solution. But until then, the rest of this chapter and the following marketing chapters are written by Steve.

The marketing approach

The different ways you can market your business are called 'channels' or 'channels to market'. A channel is way for you to reach the world, and for the world to reach you. Your ability to effectively leverage these channels to meet your vision can be the difference between business success and failure — and with 25 channels to choose from, you might be wondering where on earth should you start?

Will success come from advertising, emails, network memberships, and social selling? Or from events, exhibitions, webinars, and seminars? Or maybe organic and paid social media will deliver your desired business results?

You certainly don't have the resources to do all 25, not unless you're a very large business with infinite resources. Also, many factors influence the selection of your marketing channels, and though each has its merits, you need to pick an approach that aligns with the nature of your tech idea, your vision, objectives, and business model.

Before you review and select a channel, you must first consider what your aims are. You'll find it much easier to decide which marketing approach is best for you once you have clarity on what you want to achieve, and I recommend a structured approach to help you decide.

First, we need to start by answering three questions:

1. Where are we?

2. Where do we want to be?

3. How do we get there?

Don't just answer these questions in your head — write them down and really think about them. The answers to these questions are the first step in creating a strategic marketing plan. You'll notice that the topics covered in the book so far help you answer questions 1 and 2, but you might find that question 3 is a bit harder.

To help you get clarity on these questions, you must next consider the "7 P's of Marketing". These are the factors you need to think about to decide "How do we get there?"

The 7 P's are:

- **Product:**

  What you offer to customers including products, services, and solutions.

- **Price:**

  The price of what you offer, considering its positioning too.

- **Place:**

  The channels where your product will be purchased.

- **Process:**

  The business processes to achieve your short-, medium-, and long-term aims.

- **People:**

  The skilled resources to deliver your strategies and support you achieving your aims.

- **Physical evidence:**

  What is seen as evidence to your values, proposition, and offers.

- **Promotion:**

  The marketing channels that will be used to support the achievement of sales and marketing aims.

The last point, "promotion", is where the 25 channels come in, and you will need to consider the impact of each across one or more of the following performance criteria:

- **Awareness:**

  Building awareness of your brand, values, and products.

- **Reach:**

  The reach to your target audience where they see your messaging.

- **Traffic:**

  Attracting visitors to your various online interfaces (such as website or social channels).

- **Engagement:**

  Visitors engaging with you through your messaging and content.

- **Downloads and views:**

  Visitors downloading content, watching your videos, etc.

- **Enquiries:**

  Visitors requesting demos, assessments, further information, and conversations.

- **Leads:**

  Qualified enquiries that have Budget, Authority, Need, and are Timely (BANT).

- **Opportunities:**

  Leads that are qualified by telesales and/or sales as opportunities.

- **Orders:**

  Opportunities that are converted to orders.

Depending on the nature of your business and goals, some of these criteria may be more important to you than others. Some allow a quantitative approach, where you can measure success with data and metrics, and others may only allow a qualitative approach — where success is determined by factors that can't easily be measured but you have a general feeling that you've met them. In my opinion, it's always better to review the data to make judgements when you can, though I accept this isn't always possible.

Once you're clear on your aims (we'll get to the 25 channels in moment), you can target customers to achieve targeted aims.

I recommend you take the following journey to develop ideas around your business and marketing:

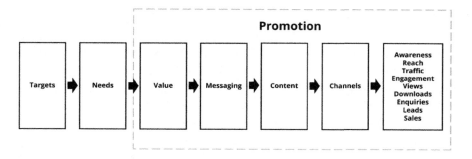

For new products, you should first define your target markets, capture target customer needs, develop your value proposition related to these needs, and ensure that the value proposition feeds the messaging. These steps should feel familiar by now.

Next, move on to the promotion stages of the journey, where you decide the content you will use to best communicate your message and the channels used to achieve your communication aims.

Once you've designed your carefully considered messaging, you can convert that messaging into desirable and engaging content that will be shared through your chosen communication channels aligned to the various stages of a customer journey.

The customer buying journey

It would be nice if we could simply identify our target customers, click our fingers, and have them buy from us. But the buying process doesn't work that way, and customers (or influencers to the purchase) first go on a journey before they buy.

The journey a customer takes will depend on the nature, importance, and value of what you're selling.

Think about the last time you bought a chocolate bar at a shop — you will likely have relied on your existing expectations from that brand to influence the decision, such as how you expected it to taste and feel. And you might not have thought much about whether to buy it or not, relying mainly on instinct and past experience. If the same brand releases a new chocolate bar, you may rely on brand experience to give you confidence that the new product will yield a similar level of experience: your enjoyment of eating the chocolate bar.

Now compare the chocolate-buying journey to the journey when you bought your last car. Even if you've purchased a certain make or model of car before, your requirements may have changed, requiring you to conduct more research into the various options. Although you may prefer a certain car manufacturer over another, you are about to spend thousands and will probably apply a lot more scrutiny to buying a new car model from the same manufacturer than a new chocolate bar from your favourite chocolate brand.

Both customers will follow this buying journey:

Though the customer journey follows the same steps with both purchase types, the time and attention at each step of the journey may shrink or grow depending on the scale of the purchasing decision to the buyer.

The higher the magnitude of the decision, the more work you'll need to put in to move the customer along each step of the journey. Why? Because you'll spend far longer exploring the options and evaluating which is best when buying a car than a chocolate bar.

It's a whopping five to 25 times more expensive to acquire a new customer than to retain an existing one. [186] And it's usually much easier to sell to existing customers than new ones as you already have a relationship with them.

What if we could leverage this effect to improve our success? What if we can sell a customer a low value item like a chocolate bar to make them five times more likely to buy a high value items like a car from us? This is what we call a "land and expand" strategy. You first land (win) the customer, then expand to your wider offering. (You can see how having an ecosystem of products or services can help with this strategy).

The marketing channel you choose to sell a chocolate bar, and then a car later will be very different to the one you'd choose if you only had a car to sell and had to sell it immediately.

For example, as a first step, 'discover' may be suited to a message in a short advert on a website, whereas a more complex and detailed 'evaluate' message may be delivered in an online webinar. The latter would be a long and more involved training course requiring a bigger time commitment from the customer and from your business to deliver it. It's unlikely that someone who doesn't know you will join your one-hour webinar if they haven't taken a lower commitment step to engage with your business first.

Here is an illustration of the steps involved in this example, where the size of each step is proportional to the number of people (potential customers) who reach the step:

Example Journey Funnel

Step 1 — Ad - Download our whitepaper
Step 2 — Email - Thank you. See our short video
Step 3 — Webinar – Learn about
Step 4 — Demo - See it in action
Step 5 — Act-now offer
Step X — Purchase

Step 1: They view your ad offering a valuable downloadable paper or eGuide, which they then download.

Step 2: They receive an email thanking them for the download and offering them a five-minute video with more detail on the business benefits. At the end of the video, they are invited to join a more detailed one-hour webinar, the offer for which is sent in an email invitation too.

Step 3: They attend a webinar. At the webinar, they are offered a free personal demonstration at your business premises.

Step 4: They attend a demonstration.

Step 5: Following the demonstration, there is a time-bound offer.

These steps may be followed by further steps, such as sharing case studies and offering telephone conversations with other customers, which eventually leads to the final step: the purchase. The size and therefore number of people who engage with each step is larger at the top and gets smaller as near the final purchase step.

As you can imagine, more people are willing to download a paper from your site for free (the equivalent of the chocolate bar) than are prepared to purchase. Many who download the paper won't be appropriate target customers either. You'll notice that the example steps look like a water funnel, which is why businesses call the particular design of steps their marketing funnel.

Some companies aim to have a small number of steps and others a larger number. More points of contact with the customer can increase your chances of making a sale, however, you need to be careful that a step doesn't become a barrier to the sale.[187] In these cases, more steps result in more customers who don't make it past the barrier, which can reduce the number of sales you make rather than help it.

You may want to adopt a hybrid approach, where customers can go down one of two funnels: a shorter one that allows you to close deals sooner if prospective clients are ready or a longer one for prospects who require more nurturing before they're ready:

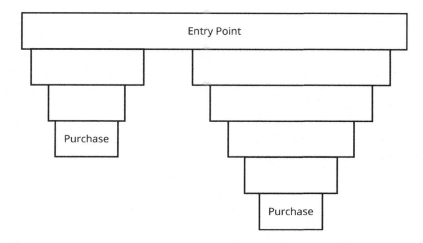

What I've discussed so far relates to funnels and marketing for a particular product offering that you sell to a prospective customer. However, I appreciate that you may be developing an internal innovation or tool.

In this case, you should think more about the user journey than the customer journey, though similar lessons apply. Your tech will likely still have an objective and a set of steps that you want users to follow on the journey to that objective. Entry points into that funnel might need to involve guiding the prospect using people and processes within your organisation rather than guiding them into the fully from marketing activity via the 25 marketing channels.

Whether you're building a product that a customer buys or an internal tool or tech solution, it's useful to understand the 25 channels and which ones your business should focus on. If you understand the marketing vision, you can create technology solutions that have longevity, and you may spot more opportunities to innovate internally if you have this wider contextual knowledge than if you don't.

Inbound vs. outbound marketing

When launching a new product for a new business, you likely won't have direct contact with the decision makers you want to reach. But if you have an established business or service offering, you may have clearer targets of who the buyer is and how you can reach them. The approach you take to marketing will depend on how well you can gather the contact information of key contacts.

If you're targeting a completely new business, how can you make them aware of you and the value you can bring when they're searching to fulfil a need or find a supplier? Or, if they're not searching and don't have an immediate need, how do you make them aware, educate them, and compel them to engage so they become aware of and compelled by the value you bring?

Inbound and outbound sales and marketing techniques are different ways of reaching the buyer based on their need and level of awareness of that need.

Outbound marketing is where you push your message out to your target audiences. This is a disruptive approach where you force-feed your messages to your target customers, hoping they're compelled by them to engage with your business.

Inbound marketing is where you market to your target audiences as they engage and interact with your business, often once they've already found you through a process of search or discovery. Inbound is often referred to as pull marketing as you're pulling customers towards your business.

To select which of the 25 marketing channels you should use, think about both the customer buying journey of your target audience and the trade-offs of inbound and outbound techniques at each step of the funnel.

The communication channels you use to support the customer journey will evolve as your brand, database, and business networks grow. For example, if your audience has no awareness of your offering, you have no target contact email address lists, aren't yet indexed in search engine results, and have no or few social media followers, then your ability to leverage these channels organically to support your go-to-market success will be limited. [188]

If you're early in your journey of building organic marketing foundations, then your future marketing strategy may involve building the supporting information and marketing infrastructure, like lists and social followers, required to make organic marketing approaches work effectively for your growth aims.

Remember the customer journey diagram? This is an expanded version of the journey we showed earlier. This version includes the baby steps a customer can take between each larger step of the journey:

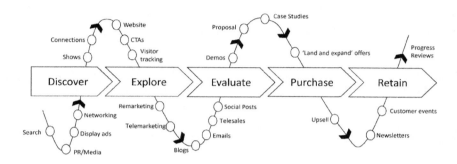

Sometimes, this process can be quick and simple, with fewer stages in the funnel behind each step and a shorter period of time at each stage, and other times it can take much longer. As with the chocolate bar and car example, what's being sold will impact the magnitude of the decision for the customer and how fast they are prepared to progress along each step.

The example journey goes like this:

- Discovery: the customer finding you

- Exploration: the customer exploring your offerings

- Evaluation: the customer evaluating your business and the factors that are important to them in their purchasing decision

- Purchase: the customer decides and takes action to buy from you

- Retention: the customer decides to stay and grow with you and continues to buy again and again (or enter into a long-term contract)

The number of customers you need to target and when to achieve your goals is another factor that impacts the channels you choose.

For example, a business with a new sophisticated business app may aim to win five new high-value customers in the first year to prove their concept before expanding their marketing strategy to target higher volume markets. The first-year journey may be based on a personalised and resource-intensive outbound marketing approach to purchasing decision-makers and influencers. It might involve a small number of carefully researched, profiled, and selected target prospects who they aim to secure those targeted clients from.

This targeted approach aligns with the MVP thinking described by Andrew earlier because the personal approach allows you to simultaneously win clients and gather the critical learning needed to adapt and improve your tech idea. However, it doesn't scale. Once you're happy that you've extracted the learning needed, switch the marketing channel you're focusing on or add new channels. This should give you more confidence that you're ready to deal with the increased business pressures that high lead and customer volumes can create.[189]

Reflect on your customer journey and think about what your prospective customers might be thinking and feeling at each stage of the journey. Consider factors like what their worries might be, what competitors they've seen, which stakeholders they need to convince, where they will be, what they'll be doing, and their priorities. You can use these factors to guide which marketing communication channels you focus on.

You should consider the available budgets, resources, and expertise when picking marketing channels. If you have limited time and money, you may only be able to target one or two channels, and not those that require large budgets (such as TV advertising). Or you may have people or tools in your organisation already specialised in one channel, such as digital advertising vs. direct forms of marketing, which supports the argument to use channels that align with your existing capabilities over those that don't.

You should also be careful not to spread your attention too thinly, as it's better to target a small number of channels extremely well than target many channels simultaneously but poorly.

As a rule of thumb, you should aim to exhaust channels that you can see are working before adding new ones, though strategically, you may ignore this rule if it takes time to develop a channel and you don't want to risk a new channel not being ready when you need it. It's fine to pick one or two marketing channels and pursue them aggressively.

Closing words

Imagine a potential customer who is actively looking for help, such as support with a need or to purchase a solution that solves their problem. Think about where they could look for this help. When did you last have a problem or need that required a solution? What did you do?

Ok, here are a few options they may try first:

1. Searching online using Google, Bing, or another search engine.
2. Searching online directories, review sites, procurement frameworks, or social media sites.
3. Asking their network, communities, friends, or others for recommendations.
4. Issuing a tender specification inviting potential suppliers to bid.
5. Seeking advice from industry analysts, consultants, influencers, or opinion leaders.

These examples are what we call search channels, and your approach to reach customers via these channels are the first five of 25 marketing channels. Clients find your business and engage with it after using a search channel or by responding to one of your outbound channels. Once engaged, you can use inbound marketing techniques to nurture visitors through the different stages of their journey.

The outbound channel choices are very diverse and include the following categories:

1. Display and video advertising
2. Social media marketing and ads
3. Social selling marketing with LinkedIn
4. Email marketing
5. Webinars and podcasts
6. Event, show, and exhibition marketing
7. Account-based marketing
8. Direct mail
9. Telemarketing
10. Partner and affiliate marketing
11. Referral marketing
12. Press relations
13. Sponsorship, public speaking, and awards
14. Industry ad placements and newsletters
15. Influencer marketing
16. Billboards and posters
17. TV and/or radio advertising
18. Point of sale marketing and merchandising
19. Conversational marketing
20. Guerrilla marketing

I've dealt with a lot of businesses over the years, and some have achieved great marketing success while others have failed. I've seen businesses with great websites and poor products perform well with their marketing, but it's rare to find a business achieve marketing success if they have a poor website.

In today's tech-enabled world, you must have an attractive and compelling website experience to support the customer journey, otherwise your customer journey funnel will look this:

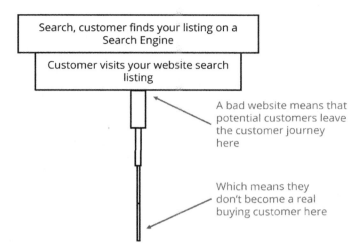

In this scenario, you could have the best possible search and outbound marketing channels in place and they could be working well, but it won't translate to results because they run into a brick wall — your ineffective website and ineffective inbound marketing.

A clear, attractive, and easy to use online experience may also be supported by an online portals, web apps, mobile apps, landing pages, or microsites.

Compelling content is required on all channels to support effective inbound marketing. Here are some examples of content that can support an effective website and your inbound marketing:[190]

- Blog content
- Email content
- Social content
- Video content
- Infographics
- Podcasts
- Guides and books

- Pain and ROI calculators
- Interactive tools and content
- Endorsements
- Case studies
- Curated content
- White papers

CHAPTER 15:
THE 25 MARKETING CHANNELS

"You can't just place a few "Buy" buttons on your website and expect your visitors to buy."
— Neil Patel

Chapter Relevance	
New Internal Tool	**New Product**
→	↑

Before you start

By this point, you should have a clear idea about the aims of your product or service and have a vision of what your customer journey could look like through the various stages of the marketing funnel. You are now ready to consider which of the 25 marketing channels are right for you!

Not every channel will be a good fit for you, so feel free to skip over it and save your energy and attention for the channels you're most interested in. For example, if you know you want to focus on digital forms of advertising, then you may not need to spend time learning about billboards and posters.

To help you cut to the right section quickly, here is a summary of the title of each marketing channel:

- Channel 1: Search engine marketing

- Channel 2: Lists, directories, and social media

- Channel 3: Networking and communities

- Channel 4: Tenders and bids

- Channel 5: Online expert sources

- Channel 6: Display advertising (on-page and in-video)

- Channel 7: Social media marketing and advertising

- Channel 8: Social selling

- Channel 9: Email marketing

- Channel 10: Webinars and podcasts

- Channel 11: Event, show, and exhibition marketing

- Channel 12: Account-based marketing (ABM)

- Channel 13: Direct Mail

- Channel 14: Telemarketing

- Channel 15: Referral marketing

- Channel 16: Strategic partners and affiliate marketing

- Channel 17: Press relations

- Channel 18: Public speaking, event sponsoring, and awards

- Channel 19: Industry ad placements and newsletters

- Channel 20: Influencer marketing

- Channel 21: Billboards and posters

- Channel 22: TV, cinema, and radio advertising

- Channel 23: Point of Sale marketing and merchandising

- Channel 24: Conversational marketing

- Channel 25: Guerrilla and viral marketing

For each channel, I will describe what it is and the impact it can have on your marketing funnel. Here are some things to keep in mind when reading each channel to decide which are right for you (and which can be ignored for now):

1. How aware the market already is of your capabilities, products, and services.

2. How many people have a specific need for that capability.

3. The number of people searching for the capability you intend to promote or for a solution to their need that the capability offers.

4. The competitiveness of the search environment and what others are using to communicate.

5. Whether you are offering a high- or low-volume product or service.

6. The complexity of your product and the difficulty in articulating its value.

7. The price of your product and the relative cost of the communication channel.

Your aims for the outcomes of your marketing will relate to:

1. Awareness and traffic

2. Engagement, downloads, and video views

3. Enquiries, leads, opportunities, and sales

Cost per action

Every marketing channel in this list has a cost associated with it. That cost might be direct, such as the amount it costs to display an advert per 1,000 users, or indirect, such as the time it takes to write a rich page of content. As with the earlier chapter about putting a value on your tech idea, you should aim to put a value on your marketing efforts.

How much you are prepared to spend to achieve a desired outcome from your marketing efforts (the *action*, such as a sale or enquiry) should be relative to the value that the action brings to you (such as the sale amount). You should avoid spending more to achieve an action than the action is worth. Conversely, don't be scared to spend more to achieve actions that are highly valuable.

For example, with digital advertising, you often pay a cost per click for your ad to be shown, and it could take between 10 and 100 clicks to get a buying customer depending on which form of advertising you choose. If each click for your product or service costs £10, then it could cost you £1,000 to acquire a customer.

Is it worth it?

If you're selling £100 handbags that people buy on a one-off basis, then maybe not, but if you're selling a £30,000 car, then spending £1,000 to make £30,000 looks a lot more attractive.

So, when considering a channel, you must always assess how much it will cost to get a customer via that channel (or combination of channels) relative to what they are worth to you (the customer lifetime value, ideally looking at profit and not revenue from the client). If you fail to check the numbers, you could end up making a loss in spending £1,000 to acquire just £100 in value, and your business won't last very long doing that! Alternatively, if you are scared to spend £1,000 to generate £10,000 (a 10x return), then you won't achieve your full potential and will hold back your growth.

For each of these channels, don't forget to ask yourself: "Does this channel effectively reach or give me a route to reach my target end customers? And is the cost to achieve my desired action for this channel worth it?"

A high cost per action doesn't necessarily mean that the channel is dead. Switching marketing channel is one way to change the cost but another is to optimise your campaigns to achieve more from a channel with the same budget. For example, if you could convert one in ten visitors to buy rather than one in twenty, then you've halved your cost per action without spending a penny more. Opportunities for optimisation exist at every stage of your funnel, and if you can improve performance by 10% at each stage, then this will result in a dramatic increase in performance throughout the whole funnel.

Most of the art of marketing is making assumptions about what will work, testing those assumptions, optimising them if they're close to performing, or dropping them to pivot to a new plan if it's clear they'll never work.

The channels

Please treat each of these 25 channels as a mini chapter, each with its own description and guidance.

Channel 1: Search engine marketing

Search engine marketing refers to the marketing techniques you can use to be found by online search engines when users are proactively searching to purchase something, get help, or find out more about topics. [191] The user's search intent may be related to your company, products, services, solutions, and content.

Researching what your prospects are searching for and aligning your offering to that search intent can bring significant results to your business: more website visitors, enquiries, leads, and sales. If done correctly, you'll achieve these positive outcomes whilst being careful where you spend and where you don't, maximising the return on your marketing budget.

If a user searches for a product or service you offer, then you want to be found on the results page of the search engine — the Search Engine Results Page or SERP.

If your offering is niche, new, or unknown, then people may not be searching for what you have to offer, and search engine marketing may not be the right channel for you. This is why it's important to research what people are searching for and how many people before making a decision.

If you can identify what people are searching for, in what volumes, and how much each customer is worth to you on average, you can decide whether it's worth your time and resources to pay to appear on the SERP for these terms.

Ways to be found

In the world of search engine marketing, there are generally two main ways to be found:

1. **Organic:** Your website pages are found by Google and rank organically, meaning they choose to show your page(s) for free because their algorithm determines that your content is both relevant to the user and high quality.

2. **Paid:** You pay to appear in search listings, often by bidding for your ads to show against certain keywords or types of search intent.

The most popular search engines have free tools to submit your business or website for listing. For example, Google has a process for registering a business profile (Google My Business) and submitting your location so you appear on their Google Maps results. Be sure to explore the options available to you across all search providers.

Appearing for free organically on search results can be difficult as it requires the search engines to trust your site and its pages, and for you to create and establish search authority. It takes time to build organic SERP performance so you may start with paid and move to organic in time for carefully chosen search terms whilst continuing to pay for others. Often, paying to appear for terms enables you to see which keywords are profitable, then you may want to invest time into trying to appear for these terms organically.

How to use search engine marketing

Remember the online model-booking application from earlier in the book that enables fashion models to connect with those looking to book models and vice versa?[192] Let's say this business is wondering what the impact of search engine marketing could be on winning new customers.

The Google Search engine has a Keyword Planning tool that allows us to see how many searches exist for different terms, called the Keyword Planner.[193] For example, this is what comes up if we use the Keyword Panner to research the search term 'book a model':

Keyword	Avg. monthly searches	Competition	Top of page bid (low range)	Top of page bid (high range)
book a model	390	High	£0.38	£1.02
model booking	20	Low	£0.49	£1.52
book a model online	10	Medium	£0.71	£1.56

Notice how even though we inputted the one search term 'book a model', the planner also returns results for similar searches?

You can see if we wanted to appear at the top of the page for 'book a model', it will cost approximately £1.02 per click. Interestingly, 'model booking' has a much higher cost per click of £1.53 even though the competition is low. Why might that be?

Can you see how the average search volume for 'model booking' is much lower than for 'book a model'? 20 monthly searches for 'model booking' compared with 390 for 'book a model'.

You pay for ad placement under an auction system, which means that higher search volume is placing downwards pressure on price because it dilutes the number of bidders and makes it more likely that a lower bid price will be accepted. The characteristics of this auction system are why it's important to consider the metrics of each keyword before deciding where to spend your search advertising budget. The most important metrics include the cost per click, the competition level, and expected search volumes.

If you have a relatively low-value or low-profit-margin product, you probably can't profitably bid for keywords that are of interest to competitors who have high-value, high-margin competing products. This is because your competitors will be willing to pay a higher amount to achieve an action than you because the value of that action is large. They have the financial firepower to outbid you.

In addition to paying to appear on the SERP, if the business wanted to perform organically, appearing for free on search results that aren't ads, they could create a new page on their website. If they call this page 'how to book a model', search engines might find it, decide it's relevant, and show it to users searching for similar terms. For this strategy to work, the page must be content-rich and relevant to the search intent of the user, especially to stand a chance of appearing for highly competitive terms.

It can be fairly easy to come up on the first page if the search competition is low. With this in mind, some companies instead choose to create several pages to target low-competitions terms. In our example, this could be achieved by creating a page called 'Model booking' given that the competition for that term is low. The business would need to identify several low-competition terms and create pages for each for this strategy to be effective.

Here is an example of a search ad for a very high search volume term: 'book taxi near me', which has over 300,000 searches every month:

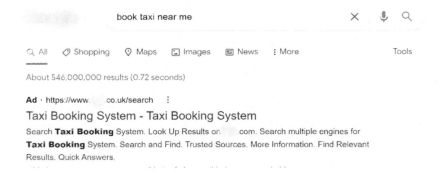

300,000 searches a month is a lot of searches! Imagine you were launching a new taxi app — if you could appear for this term cost-effectively, it could be a great way to grow your business!

There are also different places where you can bid or optimise your content to appear. For example, if your page isn't well-optimised for mobile devices, then the search engines may show your content for desktop users but not for mobile ones.

Similarly, if you're selling a tangible product, you should consider online shopping ads, which position your pages to users who are searching for shopping terms, presenting tailored information to them such as product image, prices, and calls to action, inviting the user to buy. [194]

Here is an example of an online shopping ad for the term 'buy security software': [195]

Remember you need to carefully watch how much it costs to acquire a customer relative to what you earn from that customer. In some sectors, ads can be cheap, such as £0.05 per click, but in others such as tech where product values are higher you may experience an average cost per click as high as £50!

Impact of this channel

If someone is searching to buy something, they are usually in the market to buy right now, or at least very soon. This is what we call a short sales cycle. This sales cycle can also change based on the nature of the product or service, with higher cost items requiring a longer search and discovery phase before the customer feels ready to buy.

A big benefit of paid search is that advertising campaigns can be built quickly, meaning your new product could be shown at the top of search engine results in just a few hours or days. By contrast, achieving organic search performance takes time, often many months and a lot of effort to build page authority, but for many tech companies, their organic results develop over time and can even outperform their paid search strategy. [196]

To give you a rule of thumb on the results to expect from paid search, the average click-through rate for paid search ads in the tech sector is 2.09%, and the clicks that convert into a lead or sale are 2.92%. This means you can expect a click from one in 50 impressions (the times your ad is shown to someone), and a conversion per 50 clicks or approximately one conversion per 2,500 impressions. [197]

As 99% of visitors only click on the first page of search results, if you want to achieve a click-through rate of over 5% for your organically presented search results, you will need to appear on page one of the search results and ideally in the top five. [198]

There are many tools to support your search engine marketing. [199] Here are some of the most popular free tools to get you started:

- Search Console: A free search analysis tool by Google that will help you understand and take action to improve your organic search engine marketing. [200]

- Keyword Planner: This is part of an advertising account and allows search volumes and trends to be researched. [201]

- Google Ads: If you wish to advertise, you will need to open a Google advertising account. [202]

- Google My Business: A free way of posting your business online that can be found and presented when someone searches for your company name. It allows you to be integrated with Google Maps too. [203]

- Google Analytics: A free tool that allows you to analyse website traffic, engagement, and conversions across your entire digital marketing mix. [204]

- Page Speed Insights: An online tool that allows you to test the speed and core web vitals of your website pages. [205]

- Google Trends: A tool that analyses the popularity of search queries. [206]

- Data Studio: A free visual management tool that allows the creation of a performance and tracking dashboard to monitor your sales and marketing performance. [207]

An equivalent suite of tools exists for competitor search engines. I reference Google with these resources as they are the biggest provider of search services. For example, Bing also has a suite of similar resources called Webmaster Tools. [208]

Channel 2: Lists, directories, and social media listings

Listings are places where customers go to find information. The buying behaviour is similar to search, but differs in the way that users explore and engage with the information.

Often, a purchaser will seek out lots of independent information sources to educate themselves before making a purchase. They may review the features and benefits of a product or service or evaluate the different companies available to them. Users often find this information through search engines, directories, recommendations, procurement frameworks, or from other marketing or information sources they may have seen.

Referral sources

A referral source is a resource the links to or recommends your product or service to educate or inform their users.

There is a surprising diversity of referral sources that exist! To illustrate, let's assume that our new tech product was an HR software application that provides online human resource management services. If an organisation needs such HR software, where would they look?

An online search may return HR software suppliers, but as those searching may not know what is available in the market, they may look to independent referral sites, online directories, review sites, and procurement frameworks for comparative information and recommendations.

Or let's consider directories. Many directories are free while others are chargeable or have requirements that you must fulfil to be promoted on their sites. A search on one popular directory site Capterra returns over 1,400 HR software tools that can be reviewed and compared.[209] Each one will have a list of supporting recommendations too.

In some cases, certain directories are essential to access users on a platform. For example, if you want to develop a mobile app for iOS Apple devices, then you must list that app on the official app store.

Public sector considerations

Public sector purchasers may follow different behaviours to non-public sector ones. This is because public purchasers often have existing purchasing processes that they must follow, called procurement frameworks.

These procurement frameworks are established agreements between the UK government and individual suppliers.[210] The frameworks allow government organisations to buy through the frameworks rather than them having to enter individual procurement contracts.

If a government organisation is looking for HR software, they may be more likely look to the existing digital marketplace for recommended and approved suppliers than conduct their own independent research. In other words, to reach the public sector, you may need to use public-sector specific referral sources.

If you're launching a brand-new tech product, then it's likely that you won't be on a framework — you've only just launched after all! So if you're not on the existing platforms used for public purchasing but have a tech product or service aimed at the public sector, you may be wondering how can you start and get your foot in the door?

A little trick that can help you in the short term is to identify possible partner companies that are already approved on these directories and lists, then explore whether it's possible to offer your product through them as part of their wider solution. Then you can sell through these companies into the public sector.

What makes the public sector more difficult to target via this channels is that applications to join these government supplier lists or frameworks are only open once every few years! When they do open for applications, suppliers have a short window of time to bid to become registered. If you're successful with your bid, your business will be added as an approved government supplier, and government organisations will be able to purchase from you. [211] If you're unsuccessful, then tough luck!

Social media is another way that purchasers search for information, which includes everything from recommendations to business listings and groups, most of which can be found using search keywords and hashtags. For example, a simple LinkedIn search for the hashtag #hrsoftware shows over 3,000 followers for posts in this category.

Impact of this channel

Independent references and referral sites support the discovery process for a purchaser when performing research on your company — and building a strong presence on these sites and platforms can support the trust-building and conversion process for potential customers.

There are lots of online directories that you can find on search engines to find or that you may have seen advertising in your sector. Many of these directories are free, and collectively they can provide a reasonable flow of referrals back to your website, which may lead to enquiries. Just be cautious not to aim for volume as although there are lots of directories, not all of them are good, and spending lots of time applying might be a waste.

The results that can be expected from these routes depend on the product and target audience. When potential purchasers are searching for a solution, the time to purchase is likely to be significantly shorter. Most key tech decisions are made in the first 90 days, so it's worth trying to increase your chances of reaching potential purchasers in this time window by being present in the places where they are looking. [212]

Channel 3: Networking and communities

Networking groups and communities are regular events where businesses, or businesses and consumers, get together to explore opportunities with each other. Sometimes these opportunities are direct — where business Alpha decides to buy from business Beta who they met at a networking event, and sometimes they are indirect — where business Alpha refers customer Beta to business Cappa. [213]

At these events business members and their guests meet to network and talk business. The combined relaxed and structured networking environment allows business owners to meet other businesspeople to share ideas, obtain feedback, and present what they offer.

The relaxed environment is where you mix with other attendees over coffee, saying hello and talking about whatever you want. The structured environment varies slightly but at its core, it's all about introductions to do business. Whether roundtables, where people meet in person over food or drinks, or group video calls, businesspeople give their elevator pitch to others and hand out business cards to build awareness and start relationships.

Some of the groups are highly structured and being a member means you'll be expected to pass opportunities to other businesses whilst they provide introductions and opportunities to you. These will be measured and tracked, and a lack of progress can result in your membership ending, so make sure you can afford the personal commitments required to make this channel work effectively.

Support with online activities

You'll also find a huge array of online social media communities associated with the various networking groups. Some of these online communities require approval from the heads of each group and others are more relaxed, allowing anyone to join. Conversations are conducted online within the communities and provide an opportunity to share ideas, gain feedback, and build awareness of the products you offer. The more you put into the community, the more you'll get out of it. Again, careful consideration is required to judge whether the community's members are a fit for you.

For example, on LinkedIn you have tens of thousands of 'Groups' and it's easy to open your own. With over hundreds of millions worldwide users, LinkedIn can be an excellent platform to support your networking activities, enabling you to connect with your current and future customers.

Impact of this channel

Business networking is attractive because it's local and mostly your time rather than your money. However, this blessing is also a curse as networking requires lots of time and may not be as scalable compared to marketing channels that you can grow by spending money. So, reflect carefully on the fit and likely value of the business networking groups and online communities relative to the time commitment needed to make them work.

The people who attend networking events can be owners, leaders, and decision-makers, but be warned that these groups can also be heavily orientated to salespeople selling their company's products and services with limited mutual benefit. Whilst selling is at the heart of most participants' attendance, value can also come from discussing topics with others that you would welcome input on. Often, you are looking to sell through the room by building partnerships and relationships, not selling *to* the room!

You will find that each group contains several sub-groups split by area, and each one might have a different split of companies. The profile of your target audience will determine which (if any) group is right for you.

To get the most from networking, you'll want to do your homework before each networking event. Examine attendee lists, highlight who you'd like to meet, and reach out to them in advance using social platforms such as LinkedIn saying you're looking forward to meeting them at the event. At the event, try to break through the crowd to introduce yourself to these key target people. After the event, if you haven't already done so, send them a connection request to start networking online as well as offline.

The value of business networking groups can be extremely high for some businesses, and some wouldn't even exist without such groups.

Channel 4: Tenders and bids

As you learned in "Chapter 12: How to Pick a Tech Team", tenders are a formal approach to publicly advertise your service request for suppliers to bid on.

We can't ignore the power of bidding for tenders in the commercial world. Every year, billions of pounds of UK business is done through tendering by both private and public sector companies. Although in the context of tenders as a marketing channel, you'd be the company bidding for published tenders, rather than advertising your own project work as a tender.

What to expect

The value of tender projects you can bid for may vary significantly, from the purchase of a small requirement, which may have a value of less than £1,000, to extremely high-value projects worth hundreds of millions.

Most tender opportunities will have a rigid scoring criteria that suppliers must reply to. You get points for each criteria, these get added up and compared to the other companies, then a winner is picked.

The formality of the tender can vary, and some tenders are extremely strict and bureaucratic on their requirements criteria — you either comply, partially comply, or you don't. Some tenders can be more generic, asking more open-ended qualitative questions instead.

Tender questions may also be weighted on relative importance and are comparatively scored to aid the buyer selection process. For example, let's imagine a hypothetical tender with only one question where your answer can score anything from 0 to 100%. If you were to submit a bid for this type of tender and answer in a way to score 80% against the criteria and every other bidder scored 75% or below, then your higher score would usually put you in a strong position to win the bid.

Once the provider is selected based on how they score, the successful bidder's tender response is a legally binding component of the procurement contract.

For public sector organisations or other providers spending public money, they must ensure very high levels of transparency about their purchasing decisions. This procurement transparency can be helpful if you're bidding for tenders as there will be a record of why business was placed with each supplier, and the decision is fully auditable. Many public sector organisations have a mandatory requirement to go through the formal tender process to select new suppliers.

Tenders are typically posted online on public tender portals. By exploring these tender sites, you can see what potential customers are requesting and decide whether you wish to place a bid, though it will be in a competitive environment. [214]

Your marketing activity will be aligned to support the marketing of your product and business through standout replies to tender requests, spanning from messaging and content through to presentation and supporting referenced resources.

Impact of this channel

The value of the tender-based opportunities is significant. It's estimated by UK Trade and Investment that UK public sector ICT expenditure (i.e. tech) is worth tens of billions of pounds each year. [215]

Tenders are published and easy to find but the bidding and selection process is typically bureaucratic, unpersonal, and very competitive. You may need to comply with certain standards and have industry accreditations to be able to bid, some of which involve a lot of upfront investment of time and money.

So although the prize of winning a big tender can appear attractive, be cautious in considering whether the downsides of the channel are worth the opportunities. If you chase business that you are unlikely to win, this will redirect scarce resources from places where you could win.

If you're looking for help to finding and applying for tenders, there are websites that can increase your chances of winning or guide you through the process. [216]

Channel 5: Online expert sources

It's possible to reach customers by encouraging expert sources to speak about your business and recommend you to their audience. These sources are trusted, recognised brands, so a bit of their authority comes with their recommendation, helping you to build trust and confidence.

Building authority is a powerful tactic in persuading purchasers to buy from your business, and these buyers will often seek evidence that the provider they want to buy from meets their needs and is trusted by others.

Buyers can discover high-authority brands by exploring the advice and recommendations of industry experts, analysts, consultants, influencers, and other opinion leaders' views. These include well-known bloggers, consumers, and small business tech channels, and other opinion leader sources.

For example, consumers read media sources they respect and that contain expert opinions and reviews, with each site reaching a different persona of reader. [217] Business leaders are then more likely to read different business media sources specifically aimed at them. [218]

The buyer is influenced by the complexity and importance of the purchase to them, and this influences where they choose to buy. For very large value or high complexity purchases, corporate buyers may even seek the help of specialist market research companies to inform their decision. [219] The valuable information and resources of these expert sources help purchasers compare the typically larger and established tech suppliers. If you can be listed or recommended by these sources, this can expose you to new customers or build authority in your sales and marketing process.

Impact of this channel

If you hope to make the best use of this channel, you must have a plan and process around how to market to and build relationships with industry leaders and influencers. [220] The end goal is usually to have you or your content recommended in articles, press releases, opinion pieces, ads, and advertorials (both free and paid).

Once you've built relationships with your target online expert sources, your submitted content is usually published very quickly if accepted, though acceptance is not guaranteed and many of your requests might be rejected or ignored. With this in mind, you should be careful to only spend time attempting to reach and promote through expert sources that you are confident align with your interests. You wouldn't want to dedicate a lot of time attempting to be published in an industry magazine for cats if you're selling dogs.

Channel 6: Display advertising (on-page and in-video)

Digital display adverting is where you pay to show your ads in different places, such as websites, apps, app stores, and online video providers. When clicked, these ads take visitors to your website where there's a chance that they might become a new customer or a new lead.

There are a wide range of places to display your ads. One option is to manually negotiate with site and platform owners to get access to unique deals and opportunities, but this may be difficult to scale.

An easier approach is to list your ads on an ad display network. These are platforms where you post your ads once, and it is then served in a wide range of places, such as sites, apps, and videos — anywhere that has integrated with your chosen display network. [221]

There are millions of advertisers that allow you to place ads on their website using a display network. Ever viewed a video or page on the web, then noticed ads for that thing display elsewhere on the web too? That's most likely because there's a link between the site you visited, the display network, and the sites you're visiting now. This approach of displaying ads from places you've recently visited is called 'remarketing'.

Remarketing is a great way to achieve a good bang-for-buck on your advertising spend as people who have already visited your website or app once may be more likely to buy, making them an ideal target for your ads.

Display ad formats

Display ads on these platforms are created in text, image, and/or video formats and can be displayed to targeted audiences. Examples of audience selection criteria include:

- **Remarketing**:

  By installing a special piece of code to your website that sends data to your ad network, you can remarket to your website visitors by displaying your ads to them on display, video, and social media networks. Remarketing can be generic to all website visitors or extremely targeted based on specific pages, sections, or actions they took when on your website.

- **In-market audiences**:

  This is where you display your ads to those who are in the market to purchase a specific product or service.

- **Custom audiences**:

  Here you can display your ads to niche target audiences based on what they've previously searched for (using keywords) and the websites (using URLs) they've previously visited online. For example, one custom audience may be those who have visited your competitor websites by adding the URLs of their websites to a custom audience group.

- **Similar audiences**:

  Audiences are selected of a similar profile to your website visitors and ads are displayed to them.

- **Placements**:

  This is where you define the channels you wish your ads to show on. For example, you may decide to show your ads on a popular news publication website by adding their website address to your target placements list. [222]

- **Affinity audiences:**

  This is where you display your ads to groups whose interests are likely to align with your products.

- **Customer match ads:**

  This is where lists of companies are uploaded to the advertising platform and ads are displayed to them and similar companies.

You should combine the various target criteria to reach your target audience, balancing your budget against the cost to run each type of ad. This is a similar approach to paid search ads. For example, you can narrow the criteria for where your ads are displayed and limit them to certain platforms and websites or by audience or keyword intent.

Which display ad format you choose will depend on the objectives of your campaign, such as a website ads, video ads, or banner ads. Ask yourself whether your goals are to build awareness, aid consideration, or drive action — there is a range of ad formats that enable alignment to each.

Impact of this channel

To decide whether display ads are the right channel for you, think back to the funnel discussed earlier in "Chapter 14: Marketing Principles".

- How much does it cost to put someone into the funnel?

- How many drop off on their journey through the marketing steps?

- How many buy from you at the end?

- What was the cost of that per purchase for each ad you run?

These questions will help you determine your cost per action.

The proportion of people who click your ad then go on to buy is called the awareness-to-order ratio, and it can vary greatly by the type of ad, the nature of your industry, and the products or services you sell. For example, the average click-through rate in the technology sector is 0.39%. [223] For display ad conversions, it's 0.86%. [224]

Example

Let's look at a simple example. Imagine you display ads on various sites using a display network, where you're paying the network a price for every time one of the ads is clicked.

If your average price per click for a display ad was £0.45 and the conversion rate is 0.86%, this means you'd expect to get a conversion at every 117 clicks.

Multiply this by your cost per click (117 x £0.45) and you get a cost per conversion price of £52.65.

This is an extremely crude example, and the real conversion rate would depend on many other factors, such as your offer and the journey you've designed to take people from company to customer. This may include direct forms of conversion, such as a customer buying, or indirect ones that move them one step further in the funnel, such as downloading a free resource from your site in exchange for an email address. You may be willing to spend more on ads that enable you to have a direct conversion than ones that only progress potential buyers to the next stage in the funnel.

You should consider the reach of your display campaigns. Some ads may perform, but the audience may be so narrow that it puts a ceiling on the number of clicks or engagements you can generate at any time. Think of it as a lemon versus a grapefruit — if you squeeze a lemon, you will get far less juice than if you squeeze a grapefruit. [225] Some marketing approaches are lemons, others are grapefruit. You may prefer the taste of the lemon and be prepared to pay more for it than grapefruit, but there is only so much of it available.

Research the available ad platforms to promote your product or service and see if you can find one which uniquely aligns with your offering. For example, if you plan to launch a mobile app, then you can advertise directly on the app stores to drive app downloads.

Channel 7: Social media marketing and advertising

Social media is a great way to connect with your target audience, share interesting and compelling content, and engage. Over time, prospective customers will become familiar with you and your company, your new products, and your perspective on the hot topics, innovations, or other relevant interest areas you share.

There are many social media channels to connect, share, and engage, [226] and you will find online resources that track the reach of the various social media platforms to support your decision of which platforms are right to support your business. [227]

It's important to research each channel's reach to your target audience. For example, if you wish to reach millennials (people born between 1981 and 1996), you may consider Instagram or Facebook, whereas for younger generations you may target Instagram or TikTok, and if you're looking to reach professionals, business decision-makers, or business influencers, then LinkedIn may be your best option.

How to achieve engagement

To get the most out your target social platforms:

1. Share interesting, valuable, and mixed format content.
2. Align your content to your targets' needs and create a roadmap of quality posts.
3. Use posting, monitoring, and engagement tools embedded in your site, app, or platform to aid effectiveness and productivity.
4. Be social on social media, engaging personally to build relationships.
5. Focus on constantly expanding the reach to your target audience.
6. If your new product fits, sell it using social e-commerce.
7. Track performance, then adjust and double down on what is working for you.

First, you can follow an organic strategy by posting content to extend the reach of your social media content by using well-researched high-volume hashtags and contributing to business synergist social groups or communities. You can then use paid advertising alongside your organic strategy, which will enable you to communicate with people who would otherwise be out of your network and reach. In doing so, you'll accelerate interest in your new product. Paid advertising can also give you more certainty over the expected reach of your ads compared to organic posting, which is more difficult to control.

Paid and organic content creation feed into each other as paid ads can result in increased follows, boosting engagement in non-paid content creation. Some brands even use paid ads to do nothing but build their audience and organic to reach them.

Social media formats

There are many different formats of social media advert, for example, single image ads, carousel image ads, video ads, text ads, spotlight ads, message ads, and conversation ads — and new ad formats come out all of the time!

To deliver the best results, consider which type of ad is best suited to your target audience.

You will find that the cost of advertising varies considerably depending on the social platform and whether you're trying to achieve awareness, downloads, enquiries, or sales. If you aim to achieve enquiries for your new product, focus on optimising your ads or chosen platforms to achieve the best 'cost per conversion' rather than 'cost per click'.

Here is a hypothetical example, if a click costs you £10 on LinkedIn and £1 on Facebook — surely, it's better to advertise on Facebook, right? [228] Well, maybe not. Imagine that 10 clicks on LinkedIn gets you a conversion, but for Facebook it's one for every 1,000 clicks. Even though the cost per click is 10x higher on LinkedIn than Facebook, the conversion rate is so much lower for Facebook that the cost per conversion is actually more attractive for the LinkedIn advert.

A low cost per click doesn't always mean you can expect a low cost per action.

Impact of this channel

The amount of users and the volume of daily activity across all social media platforms is staggering, however, there is also a lot of noise and distraction If you spend considerable time creating and posting content but it isn't seen by the right people as you aren't connected with them, the return on your efforts will be disappointing and you'll have wasted your time. [229]

If organic social media use is a priority for you, then building the right connections and networks on your chosen platforms is critical. And even if you plan to focus on organic, paid ads can be used in the short term to generate interest whilst you're building your profile networks.

To boost your productivity, there are also many tools that aggregate and automate posting to various social media channels. [230] Be sure to research a few to get the best results for your time.

Channel 8: Social selling

Social selling is where you use social media channels to support direct sales and prospecting activities where a person, often you or your sales team, directly reaches out to target customers on a one-on-one basis.

One of the most powerful tools to support social selling today is LinkedIn, which has hundreds of millions of registered users globally, with a focus on a professional audience. LinkedIn provides a powerful search platform and excellent business and employee database that supports prospect targeting and engagement. The depth of search into this database will depend on the level of subscription you have.

LinkedIn also provides several premium tools that make it easier to find, contact, and nurture prospects through a sales process. [231] These premium tools allow you to take full advantage of LinkedIn's search capability for both accounts and individuals, enabling detailed search and recording activity and actions against your target prospects.

Social selling search criteria

Here are some of the areas you can search for business accounts and individuals:

- For business accounts:
  - Geography
  - Industry
  - Company headcount
  - Company headcount growth
  - Department headcount
  - Department headcount growth
  - Annual revenue
  - Fortune
  - Number of followers
  - Job opportunity
  - Relationship
  - Technologies used
  - Keywords – (to include or exclude)

- For individuals
  - Keywords
  - Geography
  - Relationship
  - Industry
  - School
  - Profile language
  - First name
  - Last name
  - Seniority level
  - Years in current position

- o Years at current company

- o Function

- o Title

- o Years of experience

- o Groups

- o When became a member

- o Posted content keywords

You can add target accounts and individuals to 'Lead Lists' and 'Accounts Lists', then receive alerts from LinkedIn on interesting events related to them, prompting you with reasons to engage with them professionally and appropriately.

For example, the main alerts you can configure include:

- **Lead:** News, shares, changed job, changed role, viewed your profile, engaged with your content, accepted your connection, and recently viewed lead.

- **Account:** News, funding news, potential lead viewed your profile, account decision-makers, and senior hires at account.

Get the best results

To get the best results from your social prospecting, be careful and respectful in the way you contact and connect with them. LinkedIn has a premium messaging service called 'InMail', which provides a limited number of credits to contact people you're not connected to.

A personal and considered approach will see more of your connection requests and direct messages accepted, and you should hopefully get more replies.

Where possible, warm leads up before contacting them. You can warm up prospects by connecting with them or engaging on the platform before you reach out. Once you are connected to decision-makers and influencers, they will see your marketing content within their social feed, which will influence their awareness and perception of your business, its brand, and its value. You will also see their posts, likes, shares, and comments, giving you the opportunity to engage and build an online relationship.

Here are ten tips to be effective on LinkedIn:

1. Identify the profile attributes of your target clients.
2. Research and identify target companies using the profile attributes identified.
3. Add target companies to LinkedIn target account lists.
4. Find decision-makers and influencers in target accounts.
5. Use hashtag searches to identify possible synergist target contacts.
6. Research a valuable reason to connect.
7. Once connected, contacts will see your posts.
8. Monitor alerts for reasons to engage.
9. Appropriately engage using LinkedIn.
10. Take care in how relationships are nurtured.

There are also website tracking tools that can give a big boost to social selling strategies. [232] A big issue with contacting people on LinkedIn is knowing who the best people are to increase your chances of success, especially given the limited number of messages you're allowed to send.

You can even see which companies visit your website, and use that information to inform your sales reach out.

This visitor matching technology works by cross-referencing your visitors' IP addresses to the databases in the visitor tracking software. Though powerful, for privacy reasons you won't see the exact person who visited your site, only the company that they work for. However, you can usually cross-reference the company with LinkedIn in the tracking tools to easily identify potential prospects.

These tracking tools will also show you which pages the visitor looked at and in which order, which can prove vital in knowing what to write when you first reach out to them or what to say during the sales process.

Impact of this channel

LinkedIn's sizeable company and business database combined with its powerful search capability make it an extremely attractive social selling tool. When going to market with a new product, LinkedIn provides a quick means of connection thanks to its ability to send connection requests and InMail to communicate your message on a one-on-one basis, which can otherwise be difficult if you don't have their email or phone number.

LinkedIn and LinkedIn Sales Navigator can deliver excellent results, but don't expect them to work if you can't resource a sustained and professional approach that spans all areas from targeting to researching, monitoring to earning your connection, sharing high-value quality content, and building mutually prosperous social relationships.

Channel 9: Email marketing

Email is the workhorse of digital marketing; it has been around since near the beginning of the internet and continues to be a powerful growing channel. Most businesses include some form of email strategy in their marketing mix, and it's a widely used, effective marketing communications channel.

You should consider the following best practice tips to get the most out of your email marketing strategy:

- Segment audiences and communicate based on their needs.

- Create engaging but concise email subject lines.

- Create compelling preview text.

- Keep content concise but ensure its informative and visually engaging.

- Personalise all emails and, where appropriate, content.

- Plan the steps and calls to action you'll include in your email nurture tracks.

- Consider the appropriate time scale between emails — don't spam.

- Ensure that emails are optimised for mobile devices.

- A/B test all emails before sending them to larger audiences. This is where you run two test campaigns at the same time and proceed with the one that performs best.

If you're an established business, you will likely have built a database of email contacts, but if you're a new business, it will take time to grow your email databases. [233] Although it is possible to buy email lists, in my experience this has issues, including stale and out-of-date contacts or damaged relationships if cold outreach feels unsolicited or spammy.

You can use gated marketing content, where users must sign up to emails to access free high-value content, allowing you to build a warm, qualified opt-in list of contacts you can email. Most businesses have a series of welcome emails, in addition to regular newsletter and update emails to progress prospects through the stages of their marketing funnel.

You will work hard to secure new email contacts, and it's important to send highly relevant, informative, and valuable communication so that you keep them — and keep them engaged.

Platforms and best practice

Marketing Automation Platforms (MAPs) are another route to get the most out of your email marketing strategy. [234] These MAPs enable one-off emails and allow you to send a sequence of emails called 'nurture campaigns', which trigger on pre-defined rules actions, events, or lack of action.

For example, if you email a contact and the email isn't opened after seven days, the tool can send another one automatically. Well-designed automation campaigns help you to improve productivity and results.

If you have a sales function in your business, it's best practice to use a Customer Relationship Management system (CRM) to store prospects, leads, opportunities, and customers. Most email marketing tools can integrate with CRM systems to give your sales team information about how warm each prospect is. Some marketing tools also combine CRM and marketing automation so you only need the one system, and they may provide other rich tools such as scheduled posting of content on social media.

Impact of this channel

Email is considered by many to be *the* most important marketing channel. [235] However, if you're going to market with a new tech product as a new business, it will take time to build a quality email database that you will generate interest and results from.

Once you have a high-quality database, you can expect the following average email stats for the IT/tech sector:

- **19.5% open rate:**

    The percentage of recipients who opened the sent email.

- **2.8% click-through rate:**

    The percentage of recipients who click on a link in the email out of the total emails you've sent, regardless of whether they opened the email.

- **14.3% click-to-open rate:**

    The percentage of those who open an email and then click on a link or image within an email. [236]

Channel 10: Webinars and podcasts

Webinars are live video presentations or workshops hosted online, usually via specialist webinar software. Webinars can be used either to create demand or convert previously created demand at a later stage in your marketing funnel.

The advantage of webinars is that you can create them relatively quickly and reduce the resource overhead normally associated with larger audience face-to-face communication.

This is because there are no travel or accommodation costs, and you can reach an audience from all across the country (or world) with every webinar you host. Webinars are commonly interactive, enabling you to gain and respond to participant feedback, and they can be recorded to enable repeat plays for those who are unable to attend the live event.

Podcasts are pre-recorded or live audio clips hosted online. Podcasts have become very popular and are welcomed by business leaders who can listen whilst travelling to maximise their productivity. We have seen the introduction of audiobook companies, such as Amazon's Audible, and there is a growing appetite for audio consumption. Podcasts can help your business increase traffic, brand, reach, and leads. Listeners often subscribe, and providing your podcasts are enjoyable and informative, they will become regular listeners and loyal advocates of your brand.

Impact of this channel

If your new tech product has a detailed value proposition, webinars and podcasts are excellent marketing channels to support the evaluation phase of the company to customer journey. Webinars allow you to learn and get feedback very quickly from real customers and prospects, and they build tons of authority, enabling you or your business to establish yourself as a thought leader in your space.

Webinars and podcasts are used significantly by tech companies, in fact, about a third of all podcasts are published by technology and software companies, [237] and over 7 million people in the UK listen to podcasts each week. [238] That's one in every eight people! On average, regular podcast users listen to around seven podcasts each week, and currently podcast use is still strongly growing.

Channel 11: Event, show, and exhibition marketing

It's common for the terms events, seminars, briefings, shows, conferences, trade fairs, symposiums, and exhibitions to be used interchangeably. However, each of these terms typically means something more specific:

+ An event, seminar, or briefing will see messages shared either informally over drinks, snacks, or food or more formally through presentations and demonstrations to a target audience.

- Conferences or symposiums will see many presenters, often from different companies, share thought-provoking content and messages.

- Exhibitions are where many manufacturers or companies demonstrate their products, services, or solutions on rented floor space and an exhibition stand.

It's common for 'shows' to mix these up, so a tech show may include any of them.

Irrespective of what they're called, we'll refer to them as events, and in essence, the approach is very similar for all of them.

The three-stage approach to event prep

You can achieve great results with events if you follow a structured approach split into three stages:

1. **Pre-event preparation:**

   Ensure that the target customers are clear, your plan is clear, and you have the resources in place to achieve your goals. Make use of checklists ensure you have everything you need to host the event.

   As a rule of thumb, only 50% of those who say they will attend actually do. You usually need to invest significant effort in finding, nurturing, and chasing people to ensure the right audience attend in the right quantity. Pre-booked appointments are a must to get commitment from attendees in advance.

2. **The event:**

   Share the objectives with those attending, provide brief expectations to everyone, make sure people spend time with prospects rather than talking among the people they know, explain how interest and approval should be captured, and recognise and reward daily success.

3. **Post-event:**

   Thank everyone for attending. Hold a project review on the learning from the event. Ensure that every opportunity is followed up. Measure the leads generated through to orders to judge whether the desired return was achieved from the event spend.

Events are an excellent method to increase effectiveness if your sales process follows a direct sales approach, meaning lots of face-to-face customer visits, meetings, presentations, and demonstrations.

Instead of taking limited company resources to set up one-to-one demonstrations and presentations, events enable you to multiply your efforts so your message is presented to a larger group simultaneously. You can use your best presenters and set up more comprehensive demonstrations that would be challenging to replicate for multiple individual customer visits and demonstrations, following up with more personal one-to-one meetings for high-value prospects who showed engagement in the group events. Speaking at events also builds authority and allows you to position your business as a leader in your space.

Events enable you to establish the real meaning and professionalism of your company brand, building on what clients may have only previously seen online or perceived from telephone conversations with sales personnel.

End users value education and events can be effective in achieving this, helping you to build influence through reciprocity. When they're face-to-face, they offer the opportunity for salespeople to be present to secure the next meeting.

Impact of this channel

By focusing on the three event stages and putting strategies in place to achieve clear predefined objectives for each, events can provide an excellent communication or launch platform and a strong return on investment (ROI).

It's important to consider all related costs when calculating the event's ROI, considering both time and money invested to calculate your cost per action. You must take care to track any leads created, how they are followed up, which stages they progress to, and the end sales orders — then reflect on the total likely sales volumes for leads generated so you can accurately measure the event's success.

Channel 12: Account-based marketing

Account-based marketing (ABM) is a marketing strategy that focuses on identifying target companies and individuals and tailoring specific, targeted, and relevant content to them.

With ABM, you will aim to approach a carefully selected and profiled group of companies, called accounts, and the various decision-makers and influencers within those accounts. This strategy involves applying a high amount of energy and effort to reach and sell into a small number of very high-interest accounts.

The key to Account Based Marketing is that sales and marketing functions work together to identify the most important existing 'key target' customers and the ideal companies that, if won, will contribute to achieving business results that stretch into your medium to longer-term business goals.

Reflect on value before spending time

Before considering which key accounts to target, you should first reflect on your existing customers (if you have them), and identify which are the highest value and why. For example, it's common for 20% of your customers to account for 80% of your revenue. It's no wonder we consider these to be our key accounts!

Once you've identified your existing key accounts, think about what is similar with each. You may want to do a need-feature-benefit and persona analysis, as described earlier in the book, to give this task the attention it needs. [239] New accounts that have a similar profile to your existing key accounts have the highest future opportunity potential for you. We call these target key accounts your 'strategic key accounts', because acquiring them as customers is critical to your long-term commercial success.

Strategic key accounts aren't always high-revenue contributors today but are projected to be in the future if your ABM approach is effective. The ABM program you follow will focus on the retention and growth of existing key and strategic key accounts and winning new strategic key accounts.

Company Targeting

With Account Based Marketing, your outreach prospecting and marketing activities must be aligned to each target company. You should create a tailored growth or acquisition plan for each company and align your activities to your aims with each account. The marketing approaches you take can include the communication channels covered in this chapter, with the main difference being that each channel and its supporting content is aligned to each individual company.

For example, advertising, newsletters, and webinars are created for and targeted to the different legal entities, subsidiaries, business units, decision-makers, and influencers within one single company or group company.

But be warned as Account Based Marketing can be resource-hungry, and the total number of existing and new accounts must be kept to a level that can be effectively progressed.

If you have a prospect list that is too large, then you will spread yourself too thin if you don't narrow down your efforts. You won't achieve results if you don't allocate enough time to properly nurture each prospect. A large list can also be an indicator that you have dead leads that are wasting your time and need to be qualified out of the process. If you don't keep your target list fresh and realistic, then you risk becoming a busy fool, chasing accounts that are long gone and unlikely to yield results.

Impact of this channel

Trying to be everything to everyone dilutes your focus and understanding, resulting in a shallow proposition and poor business results.

By contrast, Account Based Marketing is all about laser-sharp focus on key strategic accounts. It's about knowing the target companies (existing and new) in immense detail, as well as the plans and supporting activity that will develop relationships and achieve strategic business aims.

To get the most out of the ABM channel, you should have a mix of accounts: some large ones that will take time to win, and some small ones that you feel are winnable much sooner. If you are unable to achieve a diverse spread of prospects, then you can combine this channel with another one that experiences shorter conversion times. This balanced approach will help you to achieve a healthy cash flow.

Channel 13: Direct Mail

Direct mail is the process of identifying the physical address of your target prospects and reaching them by sending marketing materials, such as letters, via the postal service.

As you research a prospect, it can sometimes be difficult to find their contact email or phone number, making them difficult to reach. But if you know where they work, direct mail can be an alternative way to reach them. Some businesses get creative with what they send such as posting brochures, packages or gifts, balloons, postcards, posters, or a tablet pre-loaded with a video, to name a few.

Design your messaging

Direct mail campaigns require thought about into how to get the desired message across creatively to achieve your desired aim.

You may opt to send a relatively small number of extremely personal targeted items, such as a CEO-level personally written letter to a few target accounts. Or you could opt for a more high-volume approach, sending lower-cost items to larger audiences, such as an A6 postcard with QR codes and links to online details.

Higher value items can also be part of direct mail campaigns. There have been examples of large tech companies sending high-value items to key target accounts, such as a free new model product. [240]

When sending direct mail to a senior person in a business, it's important to consider any gatekeepers, such as a personal assistant. Failure to get past the gatekeeper means that what you send doesn't reach your desired person, which can be costly due to material and postage costs, especially if sending high-value items and if it happens frequently.

Impact of this channel

Personal, low-volume, highly targeted direct mail can be effective as an introduction method where other forms of communication have failed.

The per-prospect cost of direct mail can be higher than some forms of digital advertising, however, with a well-targeted campaign, you can expect a response rate as high as around 10%. [241]

To compare the cost of your direct mail to your digital ads, you can divide the cost per reach out to the response rate to get the cost per lead. For example, if you were able to achieve a 10% response rate and it cost you £1 on average to send a letter, then your cost per lead is £10 — not too shabby!

Channel 14: Telemarketing

Telemarketing is the process of trying to reach prospects via the telephone. It's considered slightly different to the sales process as telemarketers often aren't expected to close the sale — instead passing warmed-up, qualified opportunities to the sales team to close.

However, some businesses opt to have their salespeople conduct the telemarketing too, so the approach you take can depend on the scale of your resources and the sales and marketing culture of your organisation, and it's more common to combine these roles for small companies.

Whether telemarketing is handled by marketing or sales, the effectiveness of your telemarketing will depend on how well the people making the calls are trained, supported, and enabled.

Who should do the telemarketing

Many businesses split these two activities because they have different operational and opportunity costs. [242]

Salespeople are generally more expensive to a business than telemarketers, and it's usually very inefficient to have more expensive sales resources calling 100 people a day and speaking to only a handful of them. [243] It's much better to have them focused on doing the actual selling activities!

Ideally, the right to call a target prospect is earned rather than making a cold call to someone who has no idea who you are or what you do. To earn a call, you must first engage the prospect through another form of marketing, such as email, LinkedIn, or another of the 25 channels. Earning a conversation this way will make your telemarketing more productive.

Typically, the main Key Performance Indicators (KPI) that telemarketing success are measured by are the number of appointments scheduled, how many of those appointments happen, and whether the salesperson conducting the appointments agrees that they are well-qualified to receive the call.

Have a warm-up plan

It's possible to achieve results even if you don't warm up and earn the call with the prospect – a technique called cold calling. However, this is less effective at reaching senior people as their reception team will be well trained to stop sales calls.

To get results, telemarketers must portray energy, enthusiasm, and professionalism in all of their calls. Think of how you can create a culture that helps your telemarketers ooze energy and enthusiasm into their conversations, despite the fact that their daily activities involve a lot of rejection and only a few calls are successful. Standing up to make calls can help.

It's important to create a motivational work environment for telemarketers where you celebrate success.

Be clear on value (again)

Every telemarketer on your team should articulate a consistent value proposition and be well-equipped with processes and materials to help them be effective on their calls. You can enable both sales and telemarketers with online support tools, demos, and case studies to support rewarding conversations.

Telemarketing calls should focus on the value you offer in solving people's needs and problems rather than product features, so make sure your telemarketers are trained to get this point across but also to hear before they attempt to be heard. Telephone scripts can provide frameworks and structures for the call but equally, telemarketers should come across as human, friendly, and trained to handle discussions and questions outside of a rigid script.

It's common in smaller businesses for the business owner to handle sales calls with potential customers. If you recruit a telemarketer or use a third-party agency, it's good practice to do the first few calls together with them, then get them to make the next few with you present before letting them operate autonomously on their own. The quality of the call is far more important than the number of calls made.

Earlier, we looked at how LinkedIn could be used with website visitor tracking. Telemarketing working as an integrated identification, monitoring, connection, and engagement process with these two tools will be more productive than calling a company without earning the right to a conversation. If a telemarketer blows an opportunity with a senior decision-maker, it's unlikely that they will get a second chance.

Impact of this channel

Telemarketing plays an important role within many tech companies, but it's often as part of an integrated marketing and sales process — with LinkedIn, website visitor tracking, and calls working together. It's particularly valuable in supporting the compelling articulation of more complex value propositions.

Enable your salespeople by creating events in your marketing that give them reasons to call, even if those reason are thin. It could be that you can detect which company visited your site or which users subscribed to your email list.

The number of telemarketing appointments set and then sat by a sales or telesales person per week will depend on which tech product, service, or solution you're selling and the sectors you're selling into. Some industries have decision-makers and influencers sat at their desks and are easier to contact — and in these sectors, the process is more productive.

Channel 15: Referral marketing

Referral marketing is the process of building relationships with people, businesses, or people in businesses with the goal of them passing you leads, and in some cases — high-quality, pre-qualified leads.

Referrals and business opportunity introductions from other companies can be significant contributors to achieving growth aims. However, it requires a structured approach to relationship and business development, where the relationship is mutually beneficial to both the referrer and the referee.

Referral marketing works well when combined with business networking, because with the right follow-up strategy, the people you meet at networking events can become referral partners.

Building good referral partners

Think about being a referrer from your own perspective: Why would you introduce business opportunities to other companies you know, especially if the person asked why you are recommending that company? If you refer one of your customers to use another business in your network, then a piece of your reputation goes along with that recommendation, and if the person you referred does a bad job, you might feel partly responsible.

Therefore, you must think carefully about who you ask for referrals from. It's better to have a low number of high-quality referral partners that you have a strong and trusted relationship with.

When introducing your new product, others will need to be convinced and confident in your business and your new product to recommend it. They won't want to sacrifice their relationships by introducing a new product with potential early-life bugs that lead to satisfaction issues.

The key to successful introductions

Most introductions, especially when related to higher-value opportunities, come when there is mutual benefit. The focus should be on how this can be achieved and what your value proposition to them is, and it may include:

- Keeping competitors out of their accounts by offering a wider range of solutions.

- Introducing your new product will contribute to improving their customers' satisfaction and a deeper relationship with their customers.

- An introduction fee for passing you the opportunity.

- As a result of your introduction, they are in the same or a stronger position than with their current offering.

- You may be able to pass the customer back to the referrer later in your processes to generate business for them.

For example, I run a marketing company, and Andrew runs a software development company. Sometimes Andrew will get leads for prospects who don't yet have a well-developed business strategy and so aren't ready to buy.

If Andrew loses contact with these leads, by the time they're ready to buy, they may have forgotten about his company and conduct a new search for someone to develop their tech idea for them. When Andrew refers this kind of prospect to my business, he does so because they primarily need help with their marketing, but there is also a chance that once they complete their strategic marketing plans with me, they will be ready to build their tech idea and I can pass them back to Andrew.

As you can see, the referral relationship is win-win-win for everyone involved.

Impact of this channel

Unless you're an established business, it's unlikely that you'll have existing relationships to drive plentiful referrals in the short term, making referral marketing a medium-term play.

Partnering with other companies to either create a wider solution offering or to bring additional revenue and value to their business can result in introductions to their customers, but the value to them will need to be compelling.

Business networking can be a great way to meet and build relationships with potential referral partners. If you go to a big networking event, make sure you have one-to-one meetings with people you think could become potential referral partners, and be well-equipped to explain what's in it for them.

Channel 16: Strategic partners and affiliate marketing

Strategic partners and affiliates are a levelled-up version of referral marketing.

With referral marketing, the relationships are informal — there are no written contracts or expectations. But for strategic partners and affiliates, the business relationship is deeper and more explicitly defined. It involves creating some form of contract or agreement that clearly outlines the expectations of all parties involved.

For example, if you acquired a new strategic partner for your services, then that partner might have the ability to directly sell your offering to their customers. You would then need to establish the sale price, base price, and margin the partner can make.

It's also common to have several tiers of partner (such as bronze, silver, and gold): if partners achieve a high performance, they can move to a higher tier and be eligible for additional benefits, such as increased margins or early access to new product releases and materials. Having a structured framework for incentivising partners like this is called a 'partner program'.

Franchises

As you learned earlier, a franchise is another form of strategic partnership where you give other companies the ability to set up a business using your brand and business processes and they pay you for the privilege.

Franchise relationships are common for businesses where there is a high amount of competitive advantage from the parent company, such as a strong brand, access to key locations, and procurement processes. Those buying into a franchise are looking for opportunities to invest their money in a low-risk way, as evidenced by the existing market performance of the company.

Strategic partnerships and franchising can significantly extend your reach to new potential customers. Different types of partnership include channel partners, product partnerships, system integrators, business integrators, resellers, distributors, and franchisees.

Ensure that strategic partners see benefits

For strategic partnerships to work, there must be clear value and benefit to all parties involved in the business relationship. You must ensure that your channel partners are trained, motivated, and equipped with the tools, resources, and processes need for their success.

You should be careful not to spread your efforts too thinly by aiming to get as many partners as possible. It's usually a much stronger strategy to focus on building higher quality relationships with a smaller number of partners, making sure that the incentives and processes for managing them are solid before growing the number.

I've worked a lot in partner marketing and selling over the years. Sometimes, I've deliberately cut the number of active partners to move company resources and increase our efforts in the existing key partner accounts instead, with great results.

Have a process

Regardless of the nature of the partnership, the structure and aims of the relationship should be captured in a documented partner sales and marketing plan, and there should be regular progress review meetings. Your agreement should be clear about the marketing activity that both (or either) you or the partner must complete to achieve business growth targets.

For example, marketing directly to end customers is usually easier done by the partner, equipped with your marketing training assets. But for wider campaigns, such as national advertising, it may be more appropriate for you to conduct the marketing and have a process to pass any leads generated to the partner or franchise companies to close via their own sales teams.

As you can see, making this work requires a lot of process integration and collaboration.

Affiliate marketing

Affiliate marketing is a form of partnership that still involves creating a commercial agreement between all parties involved but where the nature of the relationship is less complex and intertwined. It's more common to have simple arms-length referral-commission relationships with affiliates where they promote your product and earn a commission for each sale.

Companies will often make it very easy to sign up as an affiliate, and though the quality of affiliates still matter, there is more of a focus on volume than the other partnership types.

There are thousands of companies offering to provide affiliate opportunities and who are searching for affiliate opportunities to fulfil. [244] If you decide to sell through affiliates, make sure you focus your efforts on those whose customer base and brand align with your target audience too. Thinking back to "Chapter 7: Your Minimum Viable Product (MVP)", you should see some alignment between their need-feature-benefit-persona profile and your own.

Impact of this channel

Where there are mutual rewards, the benefits of partnering can be significant, and this is evident in the many partner programs that exist in the tech sector. If your new tech product requires a direct sales solution selling approach, partnerships may be the only way you can reach future customers who span a wide geographical area.

Having a range of affiliates marketing your new product increases your resources and reach, leading to greater awareness and hopefully sales.

Channel 17: Press relations

Press relations marketing is where you achieve results by controlling the messaging communicated by the media, such as stories in newspapers, press releases, or televised news coverage.

One strategy to leverage PR for your business is to write articles and post them to various 'news wire' websites. These news wires are frequented by journalists, and you post hoping one of them finds it, likes it, and publishes it. If you get published, the value is similar to that of advertising, where you have eyes on your brand that can raise awareness or generate leads, coupled with increased authority, as the article appears from an independent source that they trust.

Improve your success rate

To maximise your chances of being accepted and engaging readers, you should make sure that what you submit is exciting and tells a story. Write well, and if appropriate to where you are publishing, highlight the benefits and value to customers in a way that is interesting to them.

To give you an idea of what outcomes to expect, assuming you've written a strong press release that gets accepted, it usually takes a week from submission for you to see it published. It is typical to achieve around five publications by different industry sources, each with a readership of around 15,000, or approximately 75,000 in total.

This may sound a lot, but you must ask yourself how many of these readers are likely to engage with your article when yours is one of many in each publication. And of those who do engage, how many are likely to do further research or engage with your business? It may be fewer people than you expect.

You want to generate levels of awareness that lead to interest, desire, and finally, action. Awareness, Interest, Desire, Action — or AIDA for short. If you post a single press release on its own, you're unlikely to see meaningful results, and achieving AIDA requires multiple points of communication.

This Press Relations approach can be time-consuming to handle personally as a small business owner, which is why companies hire specialist press agencies who have existing strong relationships with industry media channels. Rather than posting cold through news wires, businesses leverage their established relationships to post higher quality releases more often, which get accepted more frequently and have a better chance of featuring in larger publications.

Impact of this channel

As with everything else in your marketing mix, the cost per action impacts your success. You should aim to measure the return on your money and investment, enabling you to quantify the value of press relations to your business.

With digital advertising, it's easy to track which ad campaigns deliver results, and you can even track a lead all the way from ad impression through to order. With press releases, this can be a bit more difficult, which is why it's common set up unique pages on your website for each campaign so you can track which press releases were successful.

Notice how in this book, there are lots of executeyourtechidea.com links that direct you to a different page or website? Andrew set these redirect links up so they're easier to type into a web browser, and if a web page is removed, he can change where the link leads rather than it going to a broken 404 page without needing to re-publish the book (which is possible for e-readers but much harder or impossible for print). These are called "301 Redirects", and if you have a website, you can also set up similar custom links so that when they direct to your site. These redirect links can even be set up in a way that allows you to see where redirected leads came from. [245]

Throughout my career, whenever times are tough, the value of press relations gets called into question: "Should we reduce the effort and spend in this area and direct it to other marketing channels that have demonstrable and auditable returns?"

Factors you can consider to judge the value of press relations:

- The number of placements achieved in the primary and secondary media channels.

- The growth in brand-driven organic visitors to your website or people visiting dedicated and linked website landing pages.

- Engagement data including leads and orders.

Channel 18: Public speaking, event sponsoring, and awards

I group public speaking, event sponsoring, and awards into one channel because the ways that each reach your audience are very similar.

You can promote your success across every channel you use if you're nominated or win an award to support your marketing activity. For example, if you win an award, you can shout about it on social media or via your press relations to build trust, authority, and social proof around your brand. The aim of public speaking and awards is to amplify the power of your voice and messaging everywhere.

Event sponsoring is where you pay for your product, brand or business to be promoted at a popular event. Sponsorship techniques range from sponsoring small events to major multimillion-pound deals, especially in sports where brands are looking to build their recalled values through association with the sport.

Sponsoring an event enables you to display your brand and adverts to attendees, with different amounts of exposure available depending on your level of sponsorship. This method of advertising subconsciously aligns the value of your brand to the customer's impression of the event. This positive association is why you'll often see big tech companies sponsoring big tech events: these brands want the world to think about them whenever they think about new technology.

Public speaking at these events further amplifies your message, boosting awareness, interest, and desire. Event speaking builds authority and enables you to establish yourself as a thought leader in your industry. It can also directly lead to new business leads or indirectly to opportunities that further support your business.

Impact of this channel

You can multiply the success of your existing efforts with public speaking, event sponsoring, and awards. Many businesses achieve a high number of enquiries, leads, and sales from organic engagement, and frequently the basis of this engagement is brand-based developed over time through an ongoing and sustained program of communication involving PR, sponsorships, public speaking, and awards.

Press agencies can also support you in finding events worth your time and attention to present at, increasing your chances of being selected thanks to the power of their existing relationships.

If you'd like to explore this channel without the help of a press agency, search for local exhibitions, expos, and business awards in your area. You can either pay to sponsor or speak, or if they have an award nomination process, then read the criteria to see whether you are eligible.

If you attempt to win an award, then remember that the way you write your nomination is critical — it must be well-written and powerfully present your value and achievements. People love a story, so identify a compelling and interesting narrative that makes you stand out.

Channel 19: Industry ad placements and newsletters

This channel is similar to digital advertising, except you negotiate ad placements and costs with companies directly, which may include online and offline places where your ad can be shown.

Placement sources

There are many companies who offer the ability to advertise with them and have bundles that they offer to their readers and customers, all of which you can use to promote your business. For example:

* Articles in their newsletters

- Attending their exhibitions and events

- One-off emails to their audience

- Promotion on social media

- A program of ad placements

- Opportunities for articles in their publications

- Receiving leads from their marketing

It's common for these providers to have several tiers of pricing, with each package offering more value and better opportunities for your brand or messaging to be seen.

If you're a new business that doesn't yet have an established database of contacts or emails, then marketing through established audiences in this way is an attractive means to achieve reach and score early wins.

Where to focus

Look out for companies that operate industry and technology magazines, run trade shows and exhibitions, or put on speaking conferences. These are the kinds of businesses that will allow you to advertise to their audience.

You'll find that ad placement companies have different specialisms, with some focused on running executive-level events held in various attractive locations. For example, some may specialise in placements on cruise ships, while others aim to appeal to more mass-market audiences. You should pick an ad placement channel that can show ads where your target customers will see them.

If you opt to advertise through the high-end placement companies, then they can also give you a path to speak to senior people attending the events. They may also include packages that enable you to have pre-scheduled face-to-face meetings with these high authority contacts.

The top-tier packages can get expensive, but whether this is a worthwhile strategy is relative to the value of each connection to your business. For example, for some tech businesses, a short conversation with the right senior person can result in millions of pounds' worth of deal flow, which makes the investment worth the potential return. The cost per action is high, but the account value is equally high which provides balance.

Impact of this channel

Industry ad placements area a great way to get a new product message out to a qualified existing audience.

When exploring industry ad placements, remember that the companies selling these marketing packages want to sell to as many other businesses as they can, including your competition.

Be diligent and understand the profile of customer you'll reach by paying to market to each audience. You should also consider the strength of the ad placement company, so check their audience size, engagement statistics, and case studies from past events.

Once you experiment with an ad placement channel, you can review the results to judge how many leads and conversions you've generated. You can then compare what you achieved against the cost of the ad package to decide whether it's worth your time and money to do it again.

Channel 20: Influencer marketing

An influencer is anyone who influences other people's opinions on social platforms, and influencers have sizable audiences that they regularly post and communicate to.

How to leverage influencer marketing

One approach using this channel is to become an influencer against and use the tools in this book to grow your audience; you can then promote your product or service to this established audience. [246]

Celebrity branded fragrances and branded products illustrate the power becoming an influencer brings. These low-margin items can be difficult to market and sell profitably for a new business with no audience, but if you already have millions of loyal fans and followers, then the marketing cost per-sale to reach that huge audience is close to zero. If you aren't an influence with an established audience, then you can't profitably sell £20 perfumes very easily when your marketing cost to achieve a sale is £25.

The other way to benefit from influencer marketing is to use other techniques reach the audience of other influencers. For example, you could paying an influencer to post content for endorsements or to attend your events.

Have structure

If you're marketing through existing influencers, start by identifying influencers that already appeal to your target audience and have an established following. Review which online, social, or offline platforms suit your objectives, and identify target influencers on each of those channels.

If your offering is extremely niche, then find influencers who also fill that niche, as this increases the likelihood you will be successful. Sometimes it can be better to approach a handful of up-and-coming influencers which a growing audience, rather than a single large one. These growing influencers can be more willing to collaborate and are in less demand, so you'll have more power to negotiate.

There are various types of influencer, from fashion and consumer tech through to influencers who focus entirely on categories in the business-to-business (B2B) space.

Business influencers are experts in their fields — they may be part of the public-speaking circuit, have written a business book, and be well-followed. You can explore ways to form mutually beneficial relationships with business influencers in return for support, such as building a partnership, referral, or affiliate relationships as mentioned previously.

You can also leverage these influencers' credibility by paying them for services they offer, such as them being a model for your brand, creating collaborative video content, speaking at a company event, or being a guest author.

Impact of this channel

Be cautious before engaging an influencer — make sure they're still active and their audience is engaged, as sometimes people amass large audiences and then neglect them. If you attempt to market through an influencer that has long neglected their large audience, then their posts may have reduced reach, which can significantly hinder the results you achieve. [247]

Word-of-mouth marketing is extremely powerful and is estimated to account for 13% of consumer sales with impressions contributing to five times more sales than a paid media impression. People are 90% more likely to trust and buy from a brand recommended by a friend. [248] Think of marketing through influencers as word-of-mouth marketing on steroids, allowing you to be recommended by a friend (the influencer) to their thousands or even millions of fans.

Be cautious approaching influencers who also endorse several other products or brands that compete with yours, as if it goes too far, it may dilute your message and reduce the customers' likelihood of buying. [249]

There is also a difference in the way people buy when they are alert and take time to think about the purchase compared to when they buy things in the moment without considering it. It appears that people are more likely to respond and buy based on a celebrity or influencer endorsement when they are making a quick unconsidered purchase rather than a well-considered one that they take their time with. [250] This effect makes certain types of products and service more well-suited to marketing through influencers.

Channel 21: Billboards and posters

Billboard and poster marketing is where you design and display a large-format advert on billboards or posters, either printed or digital. This is channel is a form of out-of-home advertising (OOH for short).

You can find OOH advertising when you are waiting to board an underground train on the platform, stood under a shelter waiting for a bus, on public transport, at the sides of busy roads, in shopping malls, and so on... we will commonly see digital screens, paper billboards, and posters. There are thousands of outdoor digital screens you can use to promote your business, and hundreds of millions of pounds are spent every year advertising via them. [251]

If you opt for digital billboards, then it's possible to place your advert on many of them across the country by using nationwide ad agencies. You simply make your advert, format it for different billboard and poster sizes, and agree when your campaign should run to and from. [252]

Billboards and posters can also complement your other marketing channels at different stages of the marketing pipeline. For example, when attending an exhibition, you may decide to promote the location of your stand using posters on aisles and walkways that have specific actions to get visitors to come and see you.

Impact of this channel

Billboard and poster marketing is rarely used in isolation from other forms of marketing due to its inherent difficulty to track.

In the digital world, marketers will measure impressions — the number of people who see their marketing — and this helps them reflect on their total reach and reach to their target audience.

The challenge with OOH advertising is that it's difficult to accurately estimate how many views your billboard or poster will get. [253] It's not like other forms of digital advertising where you can easily track the number of clicks and impressions.

You will need to be creative to track the performance this channel. Consider techniques that allow you to test results indirectly, such as running billboard campaigns in different regions with varied messaging for a short period. If you run several campaigns at once in different areas, then you can see which performs best and apply that strategy nationwide.

Channel 22: TV, cinema, and radio advertising

Running adverts on TV and radio can be useful if you know the audience watching at the of your advert is relevant to you. TV, cinema, and radio advertising can enable you to reach hundreds of thousands or millions of people at once at a very specific time.

The American Superbowl is a great example of an event that commands an attractive audience for advertising. A 30-second ad at prime time could be seen by around 90 million people at once, where each person is a sports fan with shared interests. [254] That's a huge punch of awareness for your product and business if it aligns with those shared interests.

Conduct audience research

Audience research is critical when selecting where to advertise. Even though the reach is wide, reaching high volumes of people costs money. If you pay to reach a large audience of people who are disinterested in your product, then you can waste a lot of money. Prices can vary wildly by channel, location, and time, so think carefully about when your target audience is likely to be watching — and when your non-target audience isn't. [255]

TV advertising bundles on niche channels for small businesses can be cheaper than you would expect, especially if the niche channel targets an audience in your area of interest. This focussed approach can see you achieve lower costs, higher targeting, and better results.

Impact of this channel

What was the last tech advert you saw or heard on the TV or radio? What was it? And why were you watching or listening?

If you see a competitor repeatedly advertising on TV, cinema, or radio, then it's probably because it's working for them. Observe how they've structured their ad and who they're aiming to reach, and see whether you can learn and copy their success.

Ideally, you should try to have your marketing message shown to your target audience many times via different channels to increase the likelihood they will buy. In the 1930s, this was called the 'rule of seven' principle, suggesting the magic number was the prospective client seeing your messaging seven times. The real number to be most effective will vary for you, your business, your customers, and the way you reach them.

Given the size of its reach, TV, cinema, and radio can be a great way to have your message shown to prospective customers repeatedly as part of their journey, making this channel a great option for your brand-building goals.

Channel 23: Point of sale marketing and merchandising

Point of sale (POS) marketing is any form of marketing that is presented to you when you're in the process of considering or making a purchase in person.

For example, your product could be in a department store that sells lots of similar products and brands. If your product is one of ten others on display, you might be willing pay to stand out from the others. A common way of doing this is to advertise your product in the exact same store that is selling your product, driving more attention to your offering over the competition.

POS marketing is relatively cost-effective and helps you to grow sales from existing marketing channels.

Variations on this channel

Types of traditional POS marketing include pull-up posters, computer screen displays, and other eye-catching presentations at the point of purchase.

Traditionally, you find this nature of marketing in a retail/shop environment, but it also applies to advertising presented to the customer at the point of purchase for digital sales too. For digital sales, you can have adverts and messaging that encourage the user to buy similar or related products as they browse, or invite the customer to make a second purchase on the thank-you page as they finish their first purchase.

Merchandising is different to POS marketing and involves branding items with your company logo and strapline, such as mugs, mouse mats, pens, etc.

Merchandising is a low-cost brand-building tool to show your messaging to customers at times when you want them to think about you. High-quality, professional merchandise is a cost-effective way to promote to your target audience.

Impact of this channel

POS marketing is where the action is, and it allows you to stand out in the right place at the right time! If you can surround the product in high-impact advertising, then you can boost your results in a low-cost and high-impact way.

Channel 24: Conversational marketing

Conversational marketing uses chat, voice assistants, and conversational Artificial Intelligence (AI) to have interactional conversations with prospective customers using text, messaging, or speech.

Digital chatbots are now convincingly human, to the point that it's sometimes difficult to distinguish whether a conversation is with a human or a machine. Some artificial intelligence researchers have even claimed, somewhat controversially, that their company AI chatbot was sentient. [256]

The use of automated chatbots in recent years, is significant. [257] Customers expect rapid answers to their questions so rather than having to call or send an email, chat allows them to get instant answers to their question(s).

Consider the User Experience (UX)

If you plan to have a chat widget on your website or in your app, then make sure you think about the user experience implications of your chosen approach. Though people expect immediate answers, they may be frustrated if they expect a human and are greeted with a bot, especially if the bot is confusing to use or provides inaccurate answers.

Alternatively, if you implement a chat feature where real humans on your team respond, then make sure responses are immediate! If you implement a live chat and can't give the prospective customer an answer quickly, it may hurt the sales process more than it helps. Customers who have a poor chat experience will feel negatively towards your brand, and you may lose customers who would have purchased otherwise. Don't underestimate the processes and staff resources you may need to dedicate to an effective live chat function — you can't just install the live chat code and think you're done.

The hybrid approach

You can also combine chatbots with live chat, where you greet the customer with a bot initially, then direct them to a a live agent only for certain questions or if the customer didn't get the answer they were looking for. This can be a good way to automatically filter non-sales enquiries, allowing your more expensive sales and marketing people to focus on the enquiries that are most likely to contribute to sales.

Well-configured AI chatbots drive productive and rewarding customer conversations for your business, and they are extremely popular in the tech space where customer enquiries are a mix of sales enquiries and support requests.

Impact of this channel

Chatbots represent a significant and growing opportunity to help you grow your business by improving the customer experience. [258]

If you can make chatbots work for your business, they may increase the volume of leads you can get from your existing marketing with very little additional spend. [259]

Try to configure any automations to use a conversational style as this usually performs better, you want the person using the chatbot to feel like they're speaking with a friend. [260]

Channel 25: Guerrilla and viral marketing

Guerrilla marketing refers to a wide range of unconventional marketing techniques available to promote your business, brand, or messaging. Guerrilla marketing requires a lot of creative and out-of-the-box thinking to spot marketing opportunities that are high-impact but low-cost. [261]

Whether the campaign is fun, provocative, or contentious, the aim is the same: to achieve an emotive reaction that rapidly spreads as people share and re-share, making the Guerrilla campaign 'go viral'.

Types of guerrilla marketing

Guerrilla marketing techniques includes publicity stunts, viral videos, viral content, stencilled graffiti, inflatables, creative counter-response ads, murals, and other unusual marketing approaches, even for more conventional marketing channels.

For example, check out this guerrilla marketing technique to raise awareness for the NHS in the UK — an entertainment venue set up neon angel wings for customers to stand in front of, take photographs, and inevitably share on social media. [262]

How many poster or billboard ads do you remember sharing lately? Probably not that many. But I'm sure you could imagine yourself taking a photo with these NHS wings! If a thousand people do the same, that's a thousand free shares of your messaging that you wouldn't have received otherwise.

Another example of combining guerrilla tactics with other marketing channels are seen at tech exhibitions. As these events involve many new products being launched, exhibitors will employ considerable creativity to stand out in an extremely busy and competitive environment. In my years, I've seen magicians, dancers, gymnasts, and virtual reality demonstrations used to attract people's attention and make certain brands the talk of the show.

Viral marketing

Viral marketing is very similar to guerrilla marketing, and involves any techniques that result in the audience sharing your product or message, often via email, messaging, or social media. Guerrilla marketing often tries to be viral, but that doesn't mean that all viral marketing techniques are guerrilla ones, or vice versa.

To demonstrate the power of viral marketing techniques, imagine if everyone who sees your marketing were to share it with two other people, then those who share it do the same, and so on, then you very quickly go from really small numbers to very large ones. 1 becomes 2, 2 becomes 4, 4 becomes 8, 8 becomes 16, and so on. Double the number of shares 25 times and your marketing will be viewed a staggering 25 million times.

Another good example of a successful viral campaign was PayPal in the 2000s. Rather than spend money on marketing to achieve the growth, PayPal instead launched a paid reward scheme to encouraged social sharing. If you invited someone to use PayPal and they signed up, you would both be rewarded with a $5 cash payment.

Yeah. Well, we started off first by offering people $20 if they opened an account. And $20 if they referred anyone. And then we dropped it to $10. And we dropped it to $5.

As the network got bigger and bigger, the value of the network itself
exceeded any sort of carrot that we could offer... And then we did a bunch
of things to decrease the friction. It's just like bacteria in a Petri dish.
So what you want to do is try to have one customer generate like two
customers. OK? Or something like that. Maybe three customers, ideally.
And then you want that to happen really fast.

And you could probably model it just like bacteria growth in a Petri dish.
And then it'll just expand very quickly until it hits the side of the Petri dish
and then it slows down.
– Elon Musk, ex-CEO of PayPal

This viral scheme was so powerful that its peak saw PayPal grow by
7–10%... per day! They went on to grow their user base to over 100
million members. This campaign worked so well that PayPal still
occasionally repeat the strategy to support new user acquisition.

Impact of this channel

High-impact, low-budget Guerrilla marketing is extremely difficult
to achieve in practice. Many attempts won't move the needle and will
fail to generate results. However, when a Guerrilla campaign works,
the impact can be significant. Guerrilla marketing techniques are a
prominent way to maximise results in competitive environments such
as business trade shows.

A word of caution: guerrilla marketing can gather significant attention,
so make sure you're encouraging the *right kind* of attention. The actions
you take to achieve positive results can equally result in negative
coverage, especially if the theme is provocative to rapidly spread a
message. You don't want to do something that both spreads rapidly *and*
damages your brand.

You can build features into your tech idea that encourage your product
or service to be shared, and shared again. You could achieve virality
with cool gimmicks that impress the user, share with friends, or
through incentives that make sharing worth their while. For example,
it's common in mobile games to reward users with perks or in-game
trophies if they share certain content. Users often want to share things

they've made or built, so giving users an outlet to express and share their creativity is another way to achieve viral results.

Or perhaps there's a way you can gamify your tech idea to encourage repeated use and sharing, such as a points system or leader board. Maybe achieving virality could be as simple as asking people to share at the right time, such as when they've completed a purchase, or at the end of a piece of content they've just read.

You will find that people like to share the things they've made. So if your tech idea allows users to be creative, then make sure you enable them to share that creativity.

CHAPTER 16:
CHOOSE YOUR MARKETING CHANNELS

"Don't push people to where you want to be; meet them where they are."
— Meghan Keaney Anderson – VP Marketing at Hubspot

Chapter Relevance	
New Internal Tool	New Product
→	↑

Well done, you've made it through my extensive list of 25 channels! You now have knowledge of the different tools available to go to market and grow your tech product, and with it your business.

As you progress, please remember that it's better to focus on a small number of channels well than a large number of channels badly.

For example, a new business with limited resources going to market with a new product may start with a small number of channels. It can then grow to use more channels over time as sales increase and the business develops the capacity to service additional channels.

How to evaluate the channels

To help you plan which channels to use in your marketing mix, the following table provides show where each marketing channel can be used in your marketing funnel or customer journey:

Channel impact on journey stage	Discover	Explore	Evaluation	Buy
Paid search ads	X			
Search engine optimisation	X			
Referral and comparison sites	X	X	X	
Review sites	X	X	X	
Online directories	X			
Newspapers and magazines	X			
Networks and communities	X	X		
Channel impact on journey stage	**Discover**	**Explore**	**Evaluation**	**Buy**
Tenders (RFI, RFQ, RFP)	X	X	X	X
Analysts, consultants, influencers	X	X	X	
Website and blog		X	X	X
Marketing content		X	X	
Case studies			X	
Display advertising	X			
Video advertising	X	X	X	
Channel impact on journey stage	**Discover**	**Explore**	**Evaluation**	**Buy**
Social media marketing	X	X		
Social selling	X	X		
Social media advertising	X	X	X	X
Email marketing	X	X		
Webinar and podcasts		X	X	
Event marketing	X	X	X	
Direct marketing	X	X		

Channel impact on journey stage	Discover	Explore	Evaluation	Buy
Telemarketing	X	X		
Telesales and sales			X	X
Referral marketing		X	X	
Press relations	X			
Sponsorship	X			
Public speaking	X	X	X	
Awards			X	
Channel impact on journey stage	**Discover**	**Explore**	**Evaluation**	**Buy**
Networking	X	X		
Tradeshows and exhibitions	X	X	X	
Industry ad placements and newsletters	X			
Influencer marketing	X			
Posters and billboards	X			
TV and/or radio advertising	X	X		
Point of sale and merchandising	X	X	X	
Conversational marketing	X	X		
Guerrilla marketing	X			
Viral marketing	X	X		

The impact-effort matrix strikes back

Which of the 25 channels should you use to achieve your specific new product and business growth aims?

Earlier, you saw the impact-effort matrix and how it can help you decide what features to build and when. This powerful tool can also help you choose your marketing priorities.

Here is an example of what this matrix might look like for your business: [263]

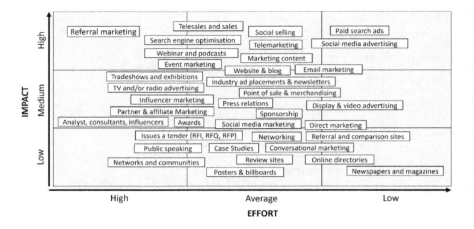

Your marketing impact-effort matrix priorities will depend on your goals and the nature of your product or service. You should complete your own version of this graph and move the position of each channel yourself.

The free resource pack includes a template file of the impact-effort matrix so you can easily drag and drop each one's channel into position based on your priorities and an assessment of each channel's value.

In addition to adding the channels to the matrix, you may also add an item for each piece of marketing or marketing content within your chosen channels. Those creating the channel campaigns are likely to be involved in creating the content too, so when scheduling your roadmap, you may wish to consider both.

Once you've plotted each channel, you can make decisions on which to progress or remove.

Consider your competition

You should think about the practicalities and level of competition to implement each channel alongside your priorities.

Low-competition channels may be less effort to reach than high-competition ones, making them an attractive place to start when testing the market for new products and services.

There will be opportunities where you will have a unique advantage, where the characteristics of your business mean you can exploit some channels in ways your competition cannot. Remember that not all channels are effective for reaching all audiences, and a channel that might work in theory is pointless if your audience doesn't engage, or even exist.

If you don't know where to start with your own marketing, then complete an impact-effort matrix by plotting all of the marketing channels you can see your competition using. Ask yourself why they use each channel, estimate the cost, and predict the likely cost per action. You can then reflect on this research to decide how these channels will work for your business.

Your competitors website, blogs, and social media posts, plus a quick Google search, will highlight many aspects of their marketing and their chosen channels. [264]

Check channel performance

As you choose your marketing channels and content strategies, set a target for each, then review how these targets relate to each other. You may find that combining two channels achieves better results than using one in isolation. Don't forget the marketing funnel we described earlier and how each channel can generate awareness, visitors, engagement, enquiries, leads, opportunities, or orders. As your business matures you should create a performance dashboard that tracks each marketing channel's actual results by month against what you originally expected and at what cost. This may sound complicated, but there are several tools mature tools that can help, some of which are free. [265]

Ensure that Analytics tools are configured to track conversions. You can also connect and pull data from other marketing and CRM tools to create a full view of your marketing and sales funnel. Once it's set up, it should be shared with all stakeholders who will can then track progress through their web browser. [266]

Closing words

Successfully launching a new product can be compared to launching a rocket into space. A successful launch requires planning, preparation, and focus on each launch stage:

- **Prelaunch:** There are many deliverables that must be ready to support a successful launch and achieve the launch objectives. Plan what they are, assign responsibilities, build them well and on time, and track progress. The items required to successfully 'take off' must be continually checked as the count down to the launch occurs.

- **The launch:** When the launch 'button' is pressed, sustained energy is essential to carry the rocket through Earth's gravity and into space. The checks must continue as the rocket progresses on its journey to break through Earth's atmosphere, just as a new product will need energy to carry it from the early adopters to the mainstream market.

- **Post-launch:** Once the rocket (the product) has arrived in space (the market), the focus shifts to reaching the destination (growth targets). The focus is on ensuring that each component (marketing campaigns, content, etc.) optimally performs to guide and accelerate the rocket to its desired destination (targeted financials). Success is only achieved when the destination is reached and the rocket successfully returns to Earth.

There is a reason why many small businesses seeking higher growth regret not investing in marketing sooner. [267] Companies often invest sizeably into developing a new product, then under-invest in marketing, meaning their product sadly fails to achieve go-to-market results. If you can forecast the resources you need to hit your objectives, then you can invest the necessary time and effort to launch your product successfully.

Key takeaways:

1. Create your strategic and marketing plan considering the 7 P's of marketing (Product, price, promotion, place, people, process, and physical evidence).

2. Define your target market, capture your customer needs, and develop your proposition.

3. Reflect on the company-to-customer journey and the steps in the discover, explore, evaluate, and purchase stages.

4. Reflect on the impact of each of the communication channels on achieving your aims, respecting the journey and the steps along the journey.

5. Use the impact-effort matrix to help you prioritise and select the communication channels that will form part of your resourced execution roadmap.

6. Research and consider your competitors' marketing when selecting your channels.

7. Create compelling content that supports your communication and journey aims.

8. Summarise your targets in visual performance dashboards and constantly track the results.

9. Constantly tweak and optimise your approach to stay on track and achieve the desired results.

CHAPTER 17:
PLAN YOUR PRODUCT ROADMAP

*"Nothing is less productive than to make more efficient
what should not be done at all."*
— *Peter Drucker*

Chapter Relevance	
New Internal Tool	**New Product**
↑	↑

Your product roadmap is the plan for your tech idea's future features.

Once your product has launched, your marketing is underway, and everything is running smoothly, you might be tempted to sit back and watch the results happen. No more work needed, right? You're done?

Probably not, and if your project is successful, then launching the first version of your tech idea is likely just the beginning of the work. It's therefore critical that you have a strategy about what to do next, and techniques to keep track of your features and priorities.

For small internal projects, it's possible to launch your idea and not come back to it very often, but for most projects, it's very rare to achieve a runaway success immediately after you launch.

Once you release your tech idea to the world, you'll begin to get feedback and learn lessons not possible in planning. Maybe customers don't use the product how you hoped or they do use it but consistently get stuck at a certain point. Maybe that point is preventing many users from performing a desired action, such as completing an order or submitting their payment information.

Even if you launch your first version and are lucky enough to benefit from immediate revenue, the story likely won't end there. With success comes more opportunities to extend your solution, implement more value to protect your market position, or further automate manual tasks.

Once your idea is out there in the real world you will inevitably spot new ways to add more value. You'll tweak, update, extend, reinvent, and adapt. Your journey will feel like it goes full circle, with idea generation, feedback, idea qualification, prioritisation, comparing effort and impact, and prepping the next version for release.

This iterative process of learning and adaptation is normal, and it's necessary to ensure that your product or service continues to add value and remains relevant in a world that's constantly changing.

"The only place where success comes before work is in the dictionary." –
Vidal Sassoon

If Mark Zuckerberg had sat back with a sigh of relief after creating Facebook (Now Meta), it would now be an outdated, controversial piece of software called FaceMash that wouldn't have made Zuckerberg over 100 billion dollars. [268] Iterative change was essential to Facebook's success.

Nearly every successful tech company maintains software that is now indistinguishable from what it was when they first began their project. If you don't continue to develop your product, it could be overtaken by a competitor or become irrelevant in your industry in just a year or two. Or maybe the internal innovation has been copied by others in your industry and so doesn't add the unique value it once did, requiring an extension.

Of course, knowing what to do next after you've been working towards a single goal for so long is tricky. You have to start learning entirely new skills, creating new goals, and resetting your drive.

In this chapter, I'll take you through the next stages of your product development after its launch, focusing on project management in the form of a product roadmap. I won't tell you what to do but instead show you ways that things are commonly done and some variations, leaving it up to you which routes you'd like to go down.

What is a roadmap?

Once you've launched your Minimum Viable Product, the product roadmap will outline your next goals and the efforts required to get there.

Every project should use a roadmap after the initial launch, unless your product is small and unlikely to need much development going forward. For example, a small piece of software to be used by a few internal team members might not need a roadmap. You might build it for a narrow process and release it, and if that process doesn't change, job done (except for a small amount of support and maintenance). But a roadmap is crucial for projects that require several extensions over time.

Before deciding on features, you need a long-term vision of what you're trying to achieve or how you want your tech to evolve. Maybe you have an idea of the kind of world you want in the future and want to be part of the driving force towards it. This vision acts as a guiding force, but it isn't a strategy; it's the lens through which you decide what should go into your strategy.

Your roadmap begins by mapping out the steps you think are required to achieve this future vision, which may help you to decide upon which business goals you should progress and in which order. As we discussed in "Chapter 5: Why Your Vision Matters", often it's easier to work backwards from this vision, defining rough objectives in reverse. As you get closer to the present date and time, these steps get more specific and can be broken down into high-level requirements with a target date, and even those can be further broken down into bite-sized actionable tasks.

Though structure and process are critical to achieve success, you need to watch the vital signs of your business, and if you need to act, change, or pivot, *then do it*. You must reflect on the vision and planned steps often, adjusting your plan as things change in your business and the market. And as the steps and priorities change, so might the requirements and tasks you decide to do next.

This approach may sound messy, but your ability to react to the environment is more important than remaining strictly bound to a unchanging plan that can become outdated.

Your roadmap isn't designed to be a steadfast plan for the next few years of development but a guideline to keep you on track. As you move along, it's likely that your roadmap will shift and evolve with market demands, knowledge of your own product, and development changes.

The basic purpose of a roadmap is to build up the core of your strategic planning process and identify where you need to invest the most and why. It helps you connect your ideas to implementation and focus on driving your product forward. With a clear product roadmap, your whole team will be aligned with the same goals and know what's coming next every step of the way.

Your vision will influence your business strategy, your strategy will influence your roadmap, and your roadmap will influence your development priorities. Then you have customers, who influence everything, often at the last minute!

Company culture

The company culture in your organisation will impact the project management approach you take to maintain your product roadmap.

Let's recap the difference between Waterfall and Agile project management methods, as we'll be expanding on these principles.

- **Waterfall:** Planning happens upfront before any development starts. The development team has a clear outline of what they need to do to create the entire product. You work in defined, large, multi-month project chunks, rather than having an ongoing project development budget.

- **Agile:** You work in small time iterations called sprints. You only set firm plans for what to do in the next sprint and not much beyond that. For example, you will decide what you're going to do for the next month, develop what you defined in the month, then reassess what you need to do next as you approach the next month. Agile allows your priorities to change leading up to a sprint.

These Waterfall and Agile methods each have very different processes and outcomes when it comes to maintaining a product roadmap.

For Waterfall projects to be successful, you must have a clear idea of the structure you want them to take before you start. This means knowing what features you plan to create, when you plan to create them, and the milestone deadlines that the design and development team need to meet — even if that development happens over several months.

As we discussed earlier, you might get more budget certainty in each with Waterfall, but you can't iterate as quickly, and you need to consider and plan large chunks of your product roadmap in iterations.

You don't want to create a project working practice that results in a slow iteration cycle speed for planning, implementing, and releasing. A slow iteration cycle speed for a released tech idea is dangerous as you can't quickly respond to customer demands.

Although in most cases I recommend switching to an Agile way of working after your first version is done, it can be fine to continue to use the Waterfall method after you've launched a small internal tool as it can be a great way to control costs. But if your solution is large, has lots of customers or end users, or is part of critical internal processes, then you should think carefully about whether the Waterfall method is the right way to manage your ongoing project requirements.

Ask yourself: Can I afford the downsides of a slow iteration cycle time? If the answer is yes, then you're fine to continue working in a Waterfall way. If the answer is no, then Agile is the way to go.

The Agile method is my preferred approach for maintaining and extending a product after the launch of the first version, and its flexibility means it's brilliant at enabling you to change your plans as you gain increased knowledge of your product and audience. This knowledge grows every day that your tech idea is in the hands of real users.

However, adopting an Agile way of working is problematic for organisations that expect cost certainty, and many companies won't sign off on Agile budgets that carry a monthly cost without explicitly stating what will be delivered for that cost. Unlike Waterfall, you don't have a fixed outline of the project's scope several months in advance.

There are some budget certainties with Agile, as you can still achieve a fixed monthly budget — it's just that you won't be able to provide certainty to other stakeholders over what exactly will be delivered each month because the Agile method by design allows lots of change in response to uncertainty.

However, this Agile monthly budget approach can be beneficial to your company culture if your business has a finance department that wants to estimate the budget for development activities across a whole year. If you know your monthly budget for each sprint, you can simply multiply that by 12 to see your annual budget and ask the finance team to approve that amount at the start of each year. Just be aware that as your tech idea grows to be more successful, you may also want to grow the rate at which you develop new features, so make sure to account for this in your plans.

You should be clear on which project management method you will to use and ensure that anyone involved in the project or that the project impacts (the project's stakeholders) are aware of the trade-offs of both Waterfall and Agile approaches.

This is why culture and budget approval processes in your organisation will often dictate which method you can use. Even if you know that Agile is the best way to work after launching your tech idea, the internal politics and decision-making processes can mean it simply isn't possible to gain approval for this way of working.

This company culture restriction on project management approaches is even more relevant in the public sector, where spending has to go out to tender and the tender requirements must be written in advance of choosing a supplier. This way of sourcing suppliers doesn't easily allow for priorities or requirements to change retrospectively. [269]

Either way, make sure you pick one approach. Don't experience the worst of both worlds by attempting to work Agile while also firmly defining all requirements upfront like you would with Waterfall!

A few years ago, I was working with a customer in the tech sector who required an important extension on a project we'd already built for them and it entailed a lot of work — around six months of development. The company typically followed the Waterfall method for most projects as it met the finance department's expectations. Finance required the managers to outline what each element of spending was for new projects with a firm timeline before it was approved. As you can see, the culture of the business was skewed one way.

The project owner had very strict completion timeframes and absolutely had to finish everything by month 7. The project owner rightly decided that the Agile method was the best approach to deliver a solution within the time constraints because it allowed us to begin work immediately and conduct planning, design, and development activities in parallel with each other. So, we put together a high-level overview of each sprint and estimated a budget for the project aligned to each of these sprints.

However, the cultural norms of the company unfortunately meant that the senior management wanted the whole project to be defined with a fixed cost before they could sign off on the budget. To make matters worse, they wanted to also put the project out to tender, which moved the start date back by a further two months. They couldn't move the target completion date back any further, so what should have been a six-month project now had to be developed in just four months!

This combination of events was a deathblow to the project, as the timeframe once possible with the Agile method was simply not possible with a Waterfall approach, especially not with the much shorter deadline.

I had to do what was right, and I couldn't morally commit to a project plan that I now felt impossible, so I had no choice but to turn down the project. I found out over nine months later that the project didn't get delivered on time and was still in development. As nothing off-the-shelf existed to implement the requirements the client wanted, they eventually had to renew their licence for their outdated system and put up with its limitations for another two years. In my view, they lost a massive opportunity to innovate, and nobody won.

The Waterfall method could have worked for this project if it was possible to have started sooner, which would have given the project the time it needed to provide structure and certainty. Agile would have worked too if the culture had allowed it, but sometimes we have to work within a company's management and process constraints.

The key takeaway here is that you need to understand how your company makes decisions before settling on which technique you'll use to manage your ongoing roadmap and future features.

If your organisation won't sign off on costs until they've fully defined a plan alongside fixed costs, do you think you'll convince them to change their way of thinking? Or will you have no option but to adjust your project approach to suit the company's approach?

Expect competing interests

As you plan your product roadmap, you'll experience lots of competing interests between people, teams, suppliers, and customers. Each of these stakeholders will have varied ideas and priorities and may want the roadmap to go in a different direction to what you think should happen. As your tech idea gets larger and more successful, it will become increasingly difficult to balance these competing interests.

User feedback is just one of many areas that will raise demands on what development activities should happen next. Competing interests will come from different people within your team or departments within your organisation. Some will align with your vision and customer feedback, while others will not.

For example, your sales and marketing team might want as many useful new features as quickly as possible so they have shiny new things to talk to clients about. Your development team might want to focus on more technical elements of the project, such as code quality and infrastructure. They may want to prioritise changes that prevent risks only they can see, such as having the right foundations to cope with a significant growth in usage without it breaking.

There will be lots of noise and forces fighting for their priorities to be your priorities. You're going to be dealing with customer demand, stakeholder demand, developer demand, and demands from anyone else who has an interest in your product.

The first way to combat this pull of forces is to elect a product owner at the start of the project who is responsible for its success and balancing these competing interests. Given that you're reading this book, you may be the best person to adopt the role of product owner!

The product owner must constantly reflect on how each new requirement or change request relates to the overall vision of the solution and business, while considering the strategic direction of the organisation.

They should have thorough knowledge of the overall vision of the product whilst considering stakeholders' and customers' demands — and being able to put the vision first. This means they need to maintain strong relationships with each of the competing stakeholders, so they can negotiate a path to success in the best interests of the business, its clients, and anyone with a vested interest in the project.

Then they must decide which order to approach each request in, which things to keep, or which to reject. The product owner role has a lot of power but also a great deal of responsibility. It takes maturity to gain the level of focus needed, but if you set the stakeholders' expectations and are clear on where the product is going, it can be done. When done well, the product owner should balance all interests. [270]

Think in stories and tasks

Once you know your vision and the steps needed to get there in a way that balances the competing interests, you can begin breaking them down into requirements and tasks in an order that reflects the true sequence of priorities.

In practice, this means building a to-do list, called a backlog. You can periodically review the backlog to decide what to do in line with your priorities, estimate the work involved, and add further details to each item about how they should be approached and implemented.

Item by item, you will add items to your backlog as they come to mind, then decide which items to implement as you approach the start of each new sprint. Once you know which items are a priority, you can plan out each one, estimate the work involved, and schedule them in the upcoming sprint. This Agile process continues indefinitely.

However, this backlog of possible features can become large for big tech projects, and big lists quickly become a nightmare to manage and keep track of. It's no good managing a large backlog in a simple to-do-list app or a Word document — these methods might be fine at the beginning, but you'll soon find yourself getting lost. Instead, you need a systemised way of recording and reporting on these upcoming actions, assigning them to versions, grouping them into larger chunks of schedulable work, and having discussions with others who need to input.

Most tech companies use specialist Agile project management software to coordinate their backlog and plan their product roadmaps. [271] I recommend that you or your development team use one of these project management tools as soon as possible if you don't already. There's often a learning curve to find your way around these tools, but the benefits are way worth it.

The to-do items in your backlog can be added to project management software as 'tickets' or 'issues', the common wording these platforms use to describe a task that needs to be done.

These tickets have lots of information attached to them, such as a description, version number, files, and comments. Tickets can be grouped together or assigned different types. For example, a new feature ticket might contain different information and input fields than one representing a bug or issue.

Groups of related tickets are called 'epics' and 'stories'. Epics are groups of requirements, and stories are groups of smaller sub-tickets that make up a single requirement and are written from the perspective of the end user. Here is an example of a story:

Example story:

"As an administrator user of the system, I want to be able to synchronise information from the app into our other systems so we can track the performance of the business and the product and automate certain parts of our workflow."

Notice how this story outlines the objectives of the requirement from the user's perspective, but it doesn't go into the specifics of how that requirement should be implemented?

You can split this into a series of small but specific sub-tasks which when completed, mean the objective of the story is met.

For example, the requirements of this story could be met if the user could download data from the system to a file in a format that can be manually imported into a third-party system. Now imagine we need to write a sub-task for this:

Example subtask:

Display a "Download" button in the header area of the reports section that when pressed, triggers a spreadsheet download.

The sheet should contain the columns: Name, Address, and Email Address in plain-text format. The customer relationship management software that we use, Salesforce, has a recommended import format that you should follow when implementing this task.

This example sub-task would sit alongside other sub-tasks also assigned to the first story. If a developer started work on the original story, they would have to complete all of these sub-tasks for the story to be considered complete.

Here is a graphic to help you to visualise how epics, stories and tasks fit together:

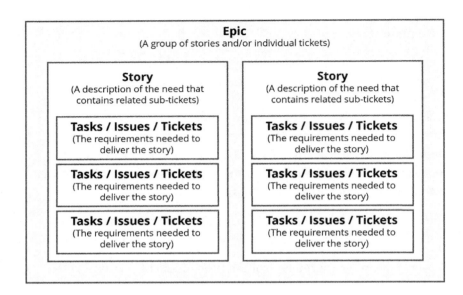

The benefit of writing the story before the sub-tasks is that the story provides useful context so you can better decide what the sub-tasks should be. Writing stories is also a quick exercise compared to writing each individual sub-task. You can optionally write some stories with no sub-tasks in advance of needing them and add detailed sub-tasks once the story becomes a priority for the next sprint.

An epic is where you group stories together in a way that makes sense to your system and your priorities. For example, you may have several stories that are part of a similar section of your system, like a reports area where several stories exist to improve its capabilities. You can group these related stories into an epic. This is useful to the development team as it's more effective to work on related features at the same time; they can get into a state of flow, and they can develop features faster in quick succession with fewer errors.

Most modern project management software should allow you create epics, stories, and sub-tasks. Once you've added your epics and stories to your backlog, you can decide which epics should form part of your next release or sprint.

You can think of each sprint as a bucket of tickets, where the size of each ticket represents how long it takes to do. Your bucket is only so big, so you must think carefully about how to fill it:

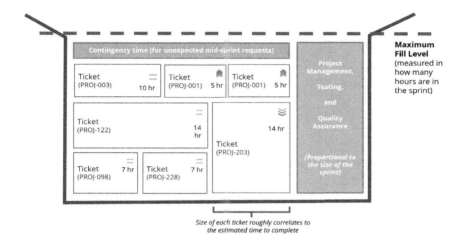

Size of each ticket roughly correlates to
the estimated time to complete

In "Chapter 10: Document Before You Implement", I explained how to create a scope and technical specification document as part of your project plan. However, the approach in this chapter is slightly different to creating a single technical specification document.

This is because the specification format I showed you earlier is usually best-suited to Waterfall projects while the story approach is better suited to Agile projects, although you can mix and match the techniques. Sometimes, the tickets you create will simply be tasks with a set of requirements on their own and won't need to be assigned to a story.

If you aren't ready to use and commit to the learning and monthly cost of project management software yet, that's ok. Similar to a technical specification, you can also plan your epics, stories and sub-tasks in a document format. Remember, you might find this approach difficult if you're working on a large projects as the document can quickly become hundreds of pages long. However, planning your stories in a document is an easy, low-barrier way to get started, and you can always copy and paste what you write into your chosen project management tools later.

To help you get started, this an example of a what a simple document version of an epic, stories, and sub-tasks plan might look like:

1. Epic Name Here

Epic Summary

Epic Requirements	Description of the goal of the epic should go here, with a summary of why the stories within the epic might be grouped in this way.
Epic Priority	HIGH / MEDIUM / LOW / LOWEST

Epic Stories

Story Name	Your Story Name Here
Story Requirements	As a [persona], I [want to], [so that].
User personas	
Story Priority	HIGH / MEDIUM / LOW / LOWEST

Subtasks

Task Name:	
Task Requirements	
Task Priority	HIGH / MEDIUM / LOW / LOWEST

Task Name:	
Task Requirements	

Notice how you start by naming your epic, summarising its requirements, and marking it with a priority? The epic then contains several stories, each with sub-tasks that carry their own description and priority.

If you'd like to use this epic, story, and sub-task template for your project, then we have a free template available on the Execute Your Tech Idea website.

Free Resource Pack

The Execute Your Tech Idea website contains further information, quick-start document templates, and other helpful free resources

executeyourtechidea.com

As your project starts to grow, and if your tech idea involves creating a digital product that will be used and extended over several years, you'll find that your backlog of tickets, stories, and epics grows with it. You can expect to gradually accumulate hundreds of tickets as your priorities shift and ideas develop.

Don't be demotivated by an ever-growing list of tickets that are pending action, as this is a normal part of developing a successful tech product. It's unlikely you will ever reach the bottom. This growing list also highlights the importance of having a prioritisation process for what you do next by reflecting on your vision and values.

As your project progresses, you need to assign time and energy to fill your backlog with actions that align with your vision and values, prioritise which tickets will be actioned next, and remove items that are no longer required.

However, you shouldn't waste time planning tickets that you might not need, and you should only add items to the list if they add value in some way and have a realistic prospect of becoming a future action. Therefore, a ticket should only sit in the backlog if it has a reasonable prospect of becoming a future action. If you no longer need it, remove it.

Remember that new information can adjust your priorities, making you re-think your decisions about what adds value. This constant change might feel messy, but it's a good thing, and it means you're making progress.

If you're outsourcing both the development and project management of your project, then your tech team may support you in writing stories, epics, and tasks and to maintain a backlog. However, even if you receive support with this, you as the project owner are responsible for the project's success and must carefully review the tickets at the beginning of each sprint. You can delegate, but don't abdicate.

The ticket workflow

Each ticket in your project management tool will have a status so you can quickly see what stage it's at. As your ticket progresses through planning to development, its status will change to reflect the stage it's at.

For example, a simple workflow for a ticket would be as follow:

1. Ticket is in the backlog.
2. Ticket is ready to be worked on.
3. Ticket is currently being worked on.
4. Ticket is now complete.

You could then scan through the status of all of the tickets in a sprint to get a rough idea of progress. If most of the tickets are now in the 'complete' stage, you know that the sprint is making good progress. As you progress with your roadmap, you'll get familiar with ticket workflows like this.

Mature project management tools allow you to create custom workflows for different kinds of ticket and task. For example, a bug may follow a different set of steps than a new feature. Development teams may also have slightly different ways of working, which will be reflected in their workflows.

Make sure you understand the workflow of your chosen tech team, as knowing the steps they follow will help you track progress.

Here's an example of a 16-step ticket workflow that I've used for some ticket types when managing tech projects. You may not need this many steps, and it might be overkill for very small projects, but hopefully it will provide some much-needed inspiration to help your tech team implement a workflow for your project. [272] Take note of the arrows between each step as they show the order of statuses that tickets transition through.

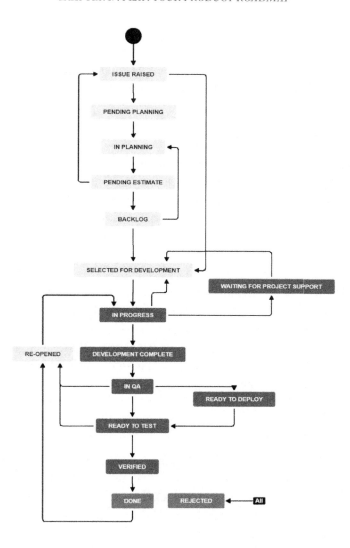

Here is what each status means: [273]

Status 1: Issue (ticket) raised

The ticket is likely to be a future action but is not yet ready to be planned.

As you manage a product, you'll often come up with ideas of things you'd like to implement or change and want to raise them when you think of them, then add detail later when appropriate.

Status 2: Pending planning

The ticket is ready to be planned. This is a signal for the project owner or responsible people in the team to add the detail to the ticket.

Status 3: In planning

The ticket is currently undergoing planning, where more detail is added to make sure the requirements are clear. Planning the ticket might follow the story approach we discussed earlier, with each ticket being given a description and definition of what it means to be done.

Status 4: Pending estimate

The ticket is now fully planned out with all of its requirements and other accompanying resources. It's ready for the design or development team to estimate how long it will take to implement and test. This is important as the time a task takes relative to the value it brings and its priority are the main influencing factors in deciding what tasks should be done and when.

Status 5: Backlog

The ticket is now fully defined, estimated, and ready to go. Tickets in the backlog will be assigned to a project version or a project sprint when they are ready to be actions. Tickets will also be assigned a priority (such as very high, high, medium, low, and very low), which will determine the order they should be worked on once they've been assigned to a version or sprint.

Status 6: Selected for development

If we're working in sprints, when a sprint begins, every ticket in the currently active sprint moves to the 'selected for development' status. This means the tasks are ready to be worked on immediately in their order of priority. The goal of the sprint is to complete all of the tickets in this queue before the sprint is over.

If the tasks are completed quicker than expected, this queue of 'selected for development' tasks can be completed before the end of the sprint. If the tasks in a sprint are completed early, then tickets in the backlog earmarked for future sprints are pulled forward into the current sprint. High-priority tickets are pulled forwards first, being careful not to disrupt the current release schedule. For example, you wouldn't pull forward a five-day sprint if only two days remain in the current sprint, as it could risk the success of the current sprint.

Status 7: In progress

This status is for tickets that are actively being worked on right now. It allows the project team to check in on development and ensure that the right things are being worked on in the correct order of priority.

Status 8: Development complete

This means the developer has completed the ticket to meet the requirements but hasn't tested it thoroughly yet. They may have performed some simple checks but not tested it altogether with the rest of the system in mind.

This status is needed as a placeholder because stories can contain several interconnected tickets. This means it isn't always possible to fully test each ticket in isolation, and a full test can only happen once all related tickets are complete. The developer will move the ticket to this status when it's complete, then test it when it's appropriate to do so.

Status 9: In QA

This status is for tickets that are currently undergoing testing by the design and/or development team. Sometimes this status is split into two steps, developer QA and project team QA if there are several layers to the testing process.

Status 10: Ready to deploy (to staging)

Once a ticket has passed testing by the development team, it's ready to be deployed to a location where the project team can test that everything works as expected.

The test location will depend on what kind of tech idea you're building. For example, for online portals, 'ready to deploy' means the work is ready for a test server, whereas for mobile apps, it means releasing a test version of the app to any test devices.

In general, we will deploy several tickets or stories of tickets at the same time for testing.

Status 11: Ready to test

Once a ticket is deployed to the testing environment, it transitions to 'ready to test', which signals to our project team that they can now review and test that the tickets meet both our quality standards and their requirements.

Status 12: Verified

Tickets move to 'verified' once the project team are happy that the ticket meets their requirements and quality standards, and is ready to be pushed to live at the end of the current sprint or release cycle. In other words, the code is 'verified' as ready to go live. All of the verified tickets are then released to the live environment at the same time.

Status 13: Done

The ticket is complete and deployed to the live environment.

I should add that there will usually be extra rounds of testing even after this live release phase that aren't outlined in this workflow, some of which can be automated. For example, before each release, you will ideally not just test the tickets from the current sprint but will test the whole system again to make sure the changes haven't caused issues with other sections of the application. This process of testing the whole system with each release is called 'regression testing'.

There will normally be an additional phase where you give the client access to the verified release and have them review and approve it before making the changes live in a real-world environment—the process of User Acceptance Testing (UAT) that we discussed earlier.

Status 14: Re-opened

If a ticket fails internal testing by the testing or project team, then gets re-opened, the development team sees it and can pick the task up and read the comments to fix the problem.

Status 15: Waiting for project support

Sometimes a ticket isn't well-defined, or the developer is blocked from being able to implement it. This status exists to catch these eventualities and prompt the project team to take action to unblock the ticket and give the development team what they need to continue working on it.

Status 16: Rejected

If a ticket is no longer required— if it's a duplicate or it has been superseded by another ticket, for example — then it gets rejected. Rejected tickets will not be actioned, though it's useful to keep a copy to report on.

Workflow structure vs. freedom

This example workflow isn't the only way to organise your project work, and there are many competing opinions about which approaches are best for software projects. Some teams prefer to adopt a low-structure approach, giving a high degree of autonomy and responsibility to their team, while others choose to be more directive, creating highly granular lists of tasks and sub-tasks.

I think it's important to maintain a balance between structure and freedom because although structure can give you control, it also reduces creativity. Sometimes it's best to tell someone exactly how you'd like something built, while other times it's best to present the problem you'd like to solve, giving your engineers freedom to use their own creativity and experience to plan a solution.

Allowing more time for creativity can result in better outcomes, and this creative activity will usually consume more time in planning but save time in execution. The challenge is that it's not always possible to estimate upfront how much creative problem-solving time might be needed. So, while it may result in faster outcomes, there's a risk that it will consume more time than you want it to, and you can't always predict which is the case in advance.

Now let's pull these concepts together. Here is a visual representation of how a planned release of your tech can consist of several sprints, each comprised of tickets. Where each ticket follows a workflow similar to the one above:

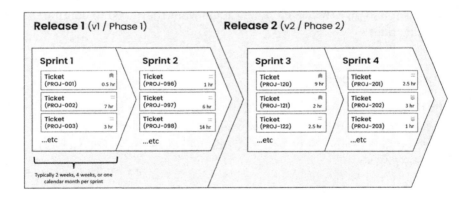

Essential technical things to do

You should always keep an eye on your product development as it moves forward. However, as a non-techy, you might not be sure what to look out for. How do you know that your developers are doing everything they should be?

To help, I've detailed some of the basic foundation tasks you should make sure happen on your project. You don't need to know all of the technical ins and outs of each of these things, but it's useful to know what they are, and why and when you need to ask for them.

Test in a staging area

Don't push your changes directly to the live environment before first performing some tests! Pushing new code changes directly into a live environment is an absolute no-no in the world of software — unless there's a valid reason. This is just lazy work and should be a huge red flag!

Adding new code directly could break things within your software, causing both security and stability issues that put your product at risk. Instead, developers should create a copy of the software on a local computer and test the code changes first. Even then, they should never push directly to live.

Your team should have a designated staging environment or a test version of the app or firmware where new code is tested and signed off. If your tech idea is used by lots of people, you may also want to consider running a live pilot test to a small subset of your users before you release to everyone else.

Automated testing

Earlier in this chapter, we touched on how we conduct a phase of regression testing as we prepare for a release to ensure that the new features don't break the old ones. Automated unit tests are programmed tests that once written, automatically regression-test elements of the system using code.

Ideally, unit tests should be written as each feature is developed, giving future developers a steadfast test that they can automatically check the functions of your app with — even if they don't understand the ins and outs of how it was built. If your developers write as they go, you'll end up with thousands of automatic tests that can check your product, saving you and future tech teams a lot of time.

Make sure you ask your development team what their plan is for unit testing. But a word of warning — writing full-coverage unit tests where every line of code in your tech project is tested at least once by a unit test will add a significant overhead to the initial build cost of your project.

However, the benefits for a large-scale software project are huge as writing these tests can have a significant positive impact on lowering the ongoing maintenance and release costs of your project. So, for small projects with constrained budgets that are unlikely to have many releases each year, your team may want to do a light-touch approach to unit tests, as it tends to be large projects that pay for themselves in the long run both in the form of saved time and better stability.

Version control

One of the most crucial and basic processes when building software is version control. Version control is essentially like a history of your development and can be used to restore old versions if anything goes wrong. It can also be used to manage different versions simultaneously, allowing two developers to collaborate and work on different elements separately and merge the versions when they're finished.

If you discover that your development team doesn't use version control tools, then a lot of questions need to be asked — this is a huge red flag! [274]

Penetration testing

Penetration testing is the process of deliberately attempting to exploit security issues with a system or application. You want to systematically try to break the tech, and if you find a way to break it, see how that broken bit can be exploited to do something nasty.

If your tech project requires a server for your database or other web-based functions, then one way to prevent security issues in your software is to identify them before your product goes live with penetration testing. These tests work by actively trying to breach your systems using known common vulnerabilities so you can see where there might be holes.

There are automated penetration testing tools that your tech team can use to run a sequence of common exploits, similar to how a virus scanner looks for dodgy code. [275] However, some penetration techniques have a real person attempt to hack into the system instead, which tends to be the more expensive option.

If you have a sizeable piece of software, penetration management can prevent data breaches in the future and secure your reputation. Running these tests can be a costly upfront expense, especially if you plan to run these tests with every software release by an independent certified penetration testing company. However, the commercial risk of you getting hacked or losing data can also be significant, so you'll need to make a judgement call about what level of approach is best for you

Code commenting

Your chosen technical team should have a process for commenting their code. Some developers aim to comment everything, while others take a pragmatic approach and add comments only where necessary. [276] The main objective of code comments is for the approach and logic behind how your tech is built to be clear and easy to understand.

Have a conversation with your team about how they plan to approach code comments and make sure you're happy with the approach suggested. But be warned, more exhaustive code-commenting methods might add additional costs to your project and seem like they add no value. The value that code comments bring is making it easier to support and maintain your project, so you should see it as an investment for the long term.

Load testing

If you have one or more servers supporting your tech, there is a limit on the number of users, devices, visits, or requests it can support before it crashes. Think of it like software running on a normal computer; once you run out of disk space, you need to upgrade your computer or get a new one to keep the software on it running. This is also true of your app infrastructure.

For example, your app may be able to deal with 10,000 users at a time, and the moment it has any more than this, it could fail. But how are you to know what your app's breaking point is? You don't want to realise you're about to hit it with little or no time to address the problem.

Sometimes, you can easily upgrade the server to cope with increased load by adding in more processors or a larger hard drive, but other times, handling the load can require time-consuming intricate work. Certain code might have to be optimised or your system may need to be split up to run on multiple servers. [277] Load testing lets you see where this limit is in advance so you have plenty of time to plan for the changes required in your product roadmap.

User Experience (UX) Research

UX research is the process of asking existing or prospective customers what they think of your changes or seeing how they react to changes without your guidance. Your tech team may be able to support you with UX research, though it's also an activity that you can do yourself if you have the capacity.

For example, as you plan which features will go into the next sprint, you may implement a process where you show the plans for these features to real users to see what they think. Sometimes, small comments and feedback from users can completely change your view about a feature, enabling you to make changes cheaply at the planning stages.

Another example of UX research comes later in the process once the features for your next release are build. Before you go live, ask several of your existing users to have a go at using the new changes. Sit over their shoulder or on a video call and watch what they do and look out for them doing something you didn't expect, such as where they get confused or use something incorrectly. If you spot an unexpected action, you may want to ask them afterwards what they were thinking at the time and use this information to improve the development or design before you release the changes into the real live environment.

Handle support requests

You must make sure you have a process for handling support requests. Even if you have the most impressive, most stable tech that ever existed, if you have users or customers, then those users and customers will raise issues.

Your priorities risk disruption if you aren't prepared for this fact, especially if you see every support request as urgent. This is especially true if a support request is shouting loudly.

Of course, you should deal with urgent support requests urgently, but please make sure you take time to reflect on their true urgency. You will find that users complain about things that are due to their own error, and you will receive some unreasonable demands.

Therefore, you should treat support requests as tickets and make sure you have a process to first qualify and prioritise a ticket yourself. Then it can then become an actionable task for your team or development team, and you can schedule this relative to your other priorities.

I recommend that you adopt a 'respond fast and action realistically' motto to handling support requests. People hate to feel ignored, but if their issue is genuinely listened to, they are far more likely to feel that the matter is in hand, even if the resolution can't be implemented immediately.

So, reply fast, set expectations on the action date or when you'll next be in touch, and communicate often, then if things change, let the person know as soon as possible.

Closing words

Make sure you plan the resource requirements for your product roadmap on if before the launch of the first version of your tech idea.

If you plan early, you can make sure you have the resources and processes in place ahead of launch rather than scrambling to resource the project retrospectively. This approach will help you to hit the ground running and make steady progress towards your future priorities.

As your tech matures and is used for real, you will notice that you begin to repeat the earlier chapters of this book. You will experience times where you must pivot and undergo periods of discovery where you re-think your vision and values. Rather than executing a new tech idea, you will seek opportunities to innovate your existing tech idea. You will need to prioritise what you do next, consider which features best align with your strategy, and which should be abandoned.

Your roadmap is an alive ever-changing beast, which sounds like chaos, but it's desirable. You must allow your plans and tech to change in line with your priorities. So listen to your customers, watch for trends in the market, and use these forces to shape what you do next.

Key takeaways:

1. Start by considering your high-level ideas for future priorities early; don't wait until your MVP is complete, and make sure you set aside a budget to continue to execute them.

2. Reflect upon your company culture and budget allocation approach, then use that to decide whether you will continue managing your project in a Waterfall or Agile manner once you've launched your MVP.

3. Use project management software to track features and requirements, and record these as stories, epics, or tasks.

4. Make sure your development team has a methodical process for recording and progressing tickets, as this will help you to keep track and organise requirements in line with your priorities, resource constraints, and budgets.

5. You can improve your tech dramatically by having a process to conduct UX research activities, and do this research in the planning phases or before your release new features.

6. Consider the priority of customer support requests in relation to all of the other tickets in your backlog and schedule accordingly.

7. Repeat the idea generation, vision and priority setting methods covered earlier in this book.

8. Listen to customer and market feedback, and use this to shape your strategy, roadmap, and next priorities.

CHAPTER 18:
SUMMARY CHECKLIST

"We all have dreams. But in order to make dreams come into reality, it takes an awful lot of determination, dedication, self-discipline, and effort."
— Jesse Owens

Chapter Relevance	
New Internal Tool	**New Product**
↑	↑

We've covered a lot — including how to come up with an idea, test its value, plan and prioritise your MVP, manage a project, pick a supplier, decide on your business model, market it, and maintain a product roadmap.

To help you remember everything you've learned, and to guide your next actions, here is a checklist of the key takeaways:

❑ Expose yourself to different sources of information and train yourself to identify solutions to problems you encounter. Keep lists until you find a tech idea that you want to start working on.

❑ Start networking to build a team of people who will be ready to give you feedback when you're ready for it. If you don't have a team yet, then pull together friends and family for feedback.

❑ As soon as you identify your potential idea, begin asking for feedback and use the responses to decide which products to pursue and which to leave on the side.

❑ Discover more about your competitive market by conducting a SWOT analysis, and use this to consider what you might do and which features you might need.

❑ Learn the basic tech terms and tasks that every supplier needs to deliver a project effectively.

❑ Start to estimate the value of your product with the following criteria: Does your idea save time? Does it save money? Does it reduce errors? Does it generate money?

❑ Research your market size to make sure there are enough people who might be interested in your product to make it worth the effort of development.

❑ Identify your target audience and their persona to start working on your value proposition (remember that different demographics within your target audience will find value in different areas of your business and respond to different messaging).

❑ Conduct a need-feature-benefit analysis to help you understand the value that your idea brings to your target market personas.

❑ Look into which revenue model will best suit your product and audience, and consider whether your chosen revenue model influences the priority and order of execution of your idea's features.

❑ Write a value proposition for your idea. If appropriate, start with each persona first before writing an all-encapsulating value proposition that applies to all of them.

❑ Prioritise features that should be included in your idea by considering the value of each persona and their corresponding value proposition. To maintain focus, you may want to begin by targeting just one persona.

❑ Work out your cash flow to predict your time-to-money (how long it will take to reach a profit).

❑ If your time-to-money is going to take too long, investigate alternative ways to generate income or investment, such as loans, start-up capital, angel investors, partners, etc.

❑ Begin plotting potential features on the impact-effort matrix to help you identify which should be the highest priority.

❑ Don't forget to ask the people on your feedback team for their opinion on your initial MVP ideas.

❑ Validate your MVP using: early-adapter landing pages, customer surveys, a pre-MVP prototype, an Excel spreadsheet prototype, etc.

❑ Create your high-level project scope and support it with wireframes if appropriate.

❑ Decide your to approach coding and development: Will you do it yourself, ask for help from an in-house tech team, source freelancers, or hire an agency?

❑ Consider which attributes of a supplier you value most: speed of delivery, quality of output, or low cost.

❑ Contact suppliers who can execute directly based on a pre-decided list of criteria.

❑ Decide which project management approach suits you or your business culture best: Waterfall or Agile.

❑ Create or outsource the creation of the wireframes for your MVP and project specification. Ideally, this will be part of a larger planning and discovery phase.

❑ If required, use the slides in the resource pack provided with this book to convince internal stakeholders that your idea is worth progressing.

❑ Engage with a supplier or tech team to build your project, entering into a contract that confirms milestones and payment terms.

❑ Understand and schedule regular project check-ins with your tech team as your project is designed and developed so you have opportunity to sign off on progress and ensure that your project is being delivered on time and to budget.

❑ Schedule time to conduct User Acceptance Testing, either in cycles if you're working Agile, or at the end of developing your MVP if you're working with a Waterfall approach.

❑ If you're launching a project, pick which of the 25 marketing channels you want to experiment with first. You may also want to consider doing this before you begin planning just in case some features are required to support your chosen marketing channel.

❑ Reflect on whether the marketing channel(s) you've chosen impact the features in your tech.

❑ Sign off on the project once it's passed UAT and agree a schedule for a live release.

❑ Assign a budget (or amount of time) required to test each marketing channel.

❑ Decide which project management software you'd like to use to maintain your product roadmap (this might be influenced by the supplier you chose).

❑ Plan a budget for ongoing support, maintenance, and a product roadmap. This budget may be designed to scale with the success of your idea.

❑ Decide if your ongoing roadmap will be managed in a Waterfall or Agile way (I recommend Agile for substantial products). And make sure you have support in your organisation and team for your chosen approach.

❑ Decide on a release cycle that matches your budgets, priorities, and chosen project management approach.

❑ Implement a ticket workflow so you can keep track of your priorities or check the workflow that your tech team would like to use.

❑ Regularly review your project priorities at set intervals, or in sprints, and review the priorities for both what needs to be planned and what is already planned.

❑ Continue to prioritise or remove new features or tickets using an impact-effort matrix.

- ❑ Continue development beyond your MVP with a product roadmap.

- ❑ Make sure you have a process in place for dealing with customer support.

- ❑ Review customer and market feedback to influence your strategy, product roadmap, and priorities.

- ❑ Remember you can find further information, quick-start documents templates and other helpful free resources on the Execute Your Tech Idea website executeyourtechidea.com

- ❑ If you found this book useful, please consider leaving a review on Amazon:

ExecuteYourTechIdea.com/r/review-book

 **Please Review on Amazon**

If you genuinely found this book useful, then it would really help me out if you could take two minutes to review it on Amazon

executeyourtechidea.com/review-book

These reviews are essential to tell the Amazon algorithm that this book adds value, and it helps others on a similar journey to find this book.

Execute Your Tech Idea was always designed to be a guide or a starting point for non-techies, not a manual to be strictly followed. The tips, advice, and knowledge here can all be filtered through your own preferences. Whilst some avenues might lead to closed doors, others may open up useable channels that you can explore as you start your development journey.

Of course, if these steps work well with your development process, then by all means use them as they are! Every piece of knowledge is based on over a decade of experience working in the tech industry, so I know that they're a solid guide to follow and can take a project from start to finish.

You might find it easier to start by taking the guide as it is and avoiding too much of your own innovation. As you learn and progress through your project, I don't doubt that new ideas will start to come naturally. As long as this book gives you the confidence to explore your potential in the world of tech development, it's done exactly what I hoped it would.

WHAT NEXT AND ABOUT THE AUTHORS

As I mentioned at the very beginning, I wrote this book based on the challenges and goals of the customers I've helped over a decade of running a software development company. [278]

With these customers in mind, I thought to myself: what is the best advice I could give everyone before they started?

As you can see from reading from this book, forming an idea is the easy part, while taking it all the way to execution requires serious focus, attention, and effort—though it is extremely worthwhile and rewarding. It also requires the business or person to have a plan to ensure that any resources supporting the launch of the project are available as and when needed, whether that's in the form of founder start-up capital, company revenues, or another source.

My business and I are based in the UK; did you know that the UK is currently the largest ICT market in Europe, with tech turnover worth several hundreds of billions of dollars? [279] You may also want to consider the opportunity to launch your tech start-up here as UK tech VC investment is third in the world after the US and China.

About the Author (Andrew)

Andrew is a UK tech entrepreneur, acrobat, powerlifter, and author.[280]

Andrews wants to see a future where technology continues to enrich people's lives and grows business productivity and wants to take steps to ensure that this kind of future will happen. Andrew aims to achieve this future by supporting or enabling businesses of all sizes to embrace technology to innovate, especially those that don't think of themselves as tech businesses. Because he sees a future where every business is a tech business.

In addition to his tech pursuits, Andrew is also a keen sports person, and won the IPF British Bench Press Championships[281] in his category three years running in 2020, 2021, and 2022, earning the privilege to represent Great Britain twice in the IPF World Bench Press Championships in Lithuania and Kazakhstan.

Find Andrew online at:
https://andrewleeward.com

Get tech execution support or learn more about Andrew's business at:
https://scorchsoft.com

And don't forget to access the free Resource pack on:
https://executeyourtechidea.com/

About the Guest Author (Steve)

Steve is a top-100 Most Influential Marketing Leader in Europe, with decades of experience working in senior management roles both nationally and internationally. [282]

Steve has helped many small, medium, and large companies to strategically plan, grow their brand, define their value, improve their communication, launch new products, win new clients, grow business with existing customers, and build high-performing marketing processes, leaders, managers, and wider teams.

Steve's experience comes from running his business, Epitomise, which provides ongoing mentoring and strategic marketing for small- and medium-sized companies, and while they don't typically do execution, they can guide you to make sure you focus on the right things in the right order.

Find Steve online at:
executeyourtechidea.com/r/epitomise

ENDNOTES

1 "Employees of firms with 2-D diversity are 45% likelier to report a growth in market share over the previous year and 70% likelier to report that the firm captured a new market." Source: How Diversity Can Drive Innovation: https://executeyourtechidea.com/r/diversity-of-thought

2 The Pareto Principle, also known as the 80/20 rule, says 80% of the results come from just 20% of the work. https://executeyourtechidea.com/r/pareto

3 Though this book is relatively long for a non-fiction, you won't find any useless and repeated filler content, just lesson after lesson to help you along your journey.

4 Often called Blue Ocean ideas.

5 Radical ideas get all the hype, but in certain contexts, it may actually be detrimental to focus on radical strategies when incremental ones are available first. Source: "Radical and Incremental Innovation Preferences in Information Technology: An Empirical Study in an Emerging Economy" https://executeyourtechidea.com/r/radical-incremental-consider

6 "The findings confirm the importance of using customer information when innovating. In line with previous studies (e.g., Evangelista, 2006), customer-related information was found to influence both radical and incremental innovation. However, collaboration with customers was found to effect only radical change. It may be that incremental innovation is largely driven by internal processes and knowledge held by employees." Source: Radical Versus Incremental Innovation: The Importance of Key Competences in Service Firms https://executeyourtechidea.com/r/radical-v-incremental-techniques

7 The Doblin Ten Types of Innovation: The Discipline of Building Breakthroughs is the culmination of 30 years of analysis and research. https://executeyourtechidea.com/r/ten-types

8   Andrew has given presentations about opportunities in tech, tech exporting, and idea generation if you're interested to hear more on this topic, for example: https://executeyourtechidea.com/r/tech-opportunities-video

9   Wikipedia article on memory allocation in the brain: https://executeyourtechidea.com/r/memoryallocation

10   @BoredElonMusk Parody Twitter profile: https://executeyourtechidea.com/r/boredelon

11   UCL (Benjamin Gardner, Phillippa Lally & Jane Wardle). Making health habitual: the psychology of 'habit-formation' and general practice https://executeyourtechidea.com/r/habitform

12   Cornell University ILR School: Jennifer S. Mueller, Shimul Melwani, Jack A. Goncalo – "The Bias Against Creativity: Why People Desire But Reject Creative Idea" – https://executeyourtechidea.com/r/creativebias

13   "The biggest risk is not taking any risk. In a world that's changing really quickly, the only strategy that is guaranteed to fail is not taking risks." – Mark Zuckerberg, Facebook Founder and CEO

14   LinkedIn social media platform for professionals: https://executeyourtechidea.com/r/linkedin

15   Median average (not mean or mode).

16   PWE Research into Facebook. 39% of adult Facebook users have between 1 and 100 Facebook friends, 23% have 101-250 friends, 20% have 251-500 friends, and 15% have more than 500 friends: https://executeyourtechidea.com/r/facebookstats

17   Popular networking groups include BNI https://executeyourtechidea.com/r/bni , Chamber of Commerce https://executeyourtechidea.com/r/chamber, and BOB https://executeyourtechidea.com/r/bob, though there are many others. Online platforms also exist for finding events such as Meetup https://executeyourtechidea.com/r/meetup

18   This is because attending events where you speak to people in-person allows you to find out their experiences and challenges today. Other forms of written media may be out of date, or less contextual to the problems of a specific person or business.

19   If you'd like to become a skilled networker and connection maker, then check out *The Clever Connector* by Ali Scarlett and Lucio Buffalmano: https://executeyourtechidea.com/r/cleverbook

20   A word of warning on online courses. Some online courses are relatively cost-effective (a few hundred pounds for a day or so of material) and provide a lot of value. However, there is an active space of practitioners offering extortionately priced courses by selling the vision of entrepreneurship. This class of online courses are promoted by Contrepreneurs (Con entrepreneurs). So, look out for promises of unrealistic results for courses that cost thousands of pounds (often $1,997). This doesn't mean all thousand-pound courses are a rip-off, but please do your due diligence before spending too much money.

21   For example, sources like eConsultancy provide rich business data and insights in your sector for a monthly fee.

22   I know 1% * 70 = 70%, I'm compounding the 1% multiplier day by day, which is $(1.01\ ^\wedge\ 70) = 200\%$.

23   Research on brainstorming effectiveness and best practices, Scott G. Isaksen and John P, Gaulin: https://executeyourtechidea.com/r/brainstormresearch

24   Credit to PayPal billionaire Peter Theil for the quote when being interviewed by Tim Ferris on the *Tim Ferris show*.

25   A Tom Ormerod study on the effects of sleep on problem-solving abilities: https://executeyourtechidea.com/r/sleepproblem

26   The frame of the game: Loss-framing increases dishonest behavior. Schindler & Pfattheicher, 2016: https://executeyourtechidea.com/r/lossframe

27   Anomalies: The Endowment Effect, Loss Aversion, and Status Quo Bias. Kahneman et al., 1991: https://executeyourtechidea.com/r/endowment

28   For superiority: The human desire for status: https://executeyourtechidea.com/r/desirestatus

29   Network effects are a driver for competitive advantage. For example, once you have a lot of active connected users, they are

the value and it's difficult for your competitors to quickly have the same size active user base.

30   United Nations list of major global issues:
      https://executeyourtechidea.com/r/globalissues

31   For example, Starling and Monzo are examples of challenger banks.

32   "Just think: The challenges of the disengaged mind":
      https://executeyourtechidea.com/r/disengagedmind

33   Netflix Subscriber Growth 2x Expectations; Good News Or Peak?:
      https://executeyourtechidea.com/r/flixgrowth

34   In 2021, there were over 1.6 billion WhatsApp users, with more than 65 billion messages sent every day, which has been achieved by great execution in combination with great value:
      https://executeyourtechidea.com/r/whatsappstats

35   For example SWOT, PESLE, Scenario Planning, and Porters Five Forces.

36   Blockbuster was a video store that operated across the USA and the rest of the world. It famously had an offer to purchase Netflix for $50m and declined. Blockbuster filed for bankruptcy in September 2010, proof that the old model for renting video content is dead. https://executeyourtechidea.com/r/blockbustercase

37   I know I've quoted Elon a lot so far, but you must admit, he does have some fantastic quotes!

38   SurveyMonkey is free for up to 100 responses (Jan 2017):
      https://executeyourtechidea.com/r/surveymonkey
      Twitter polls are another way to run a simple survey:
      https://executeyourtechidea.com/r/twitterpolls

39   https://executeyourtechidea.com/r/reddit

40   Twitter advanced search tools:
      https://executeyourtechidea.com/r/twitteradvanced

41   Such as by using Google Alerts:
      https://executeyourtechidea.com/r/galerts

42   Google Meet, Zoom, Teams, to name a few.

43   Manager tools on having effective one to ones:
      https://executeyourtechidea.com/r/mt121

44   Or a distributed monolith, where your system consists of several
      larger services interoperating.

45   If you find all of these concepts interesting and want to go slightly
      deeper, then I'd recommend the Domain of Science "Map of
      Computer Science video on YouTube.
      https://executeyourtechidea.com/r/map-of-computer-science
      It's a few years old now, but it's only 11 minutes long and offers a
      great overview of what computers can do, which might inspire
      your thinking.

46   Candy Crush Statistics:
      https://executeyourtechidea.com/r/candy-stats

47   Candy Crush revenue:
      https://executeyourtechidea.com/r/candy-crush-billion

48   The 100k Club: Most popular subscription news websites in the
      world: https://executeyourtechidea.com/r/100k-club

49   Which earned them $65 million in January 2021 alone. Source: most
      popular dating apps worldwide:
      https://executeyourtechidea.com/r/popular-dating

50   Tinder dating app example of pricing options:
      https://executeyourtechidea.com/r/tinder-app

51   Examples of SaaS include Slack, Tender Rocket, cloud-based
      Microsoft Office 365, and Google Apps (including Google Drive).
      There's also DocuSign, which lets you sign digital contracts on the
      go, Zendesk, which optimises customer service for businesses, and
      Dropbox, which lets users store and edit documents across a range
      of devices, automatically syncing changes.

52   App store study:
      https://executeyourtechidea.com/r/app-store-study

53   Apple App Store policy on selling data:
      https://executeyourtechidea.com/r/selling-policy-apple

54   Google Android policy on selling data:
      https://executeyourtechidea.com/r/selling-policy-android

55   iOS users are opting out of ad tracking:
      https://executeyourtechidea.com/r/ad-tracking-opt-out

56   Xero accounting software:
      https://executeyourtechidea.com/r/xero

57   This is a speculative example and is not based on real data
      from Xero.

58   Mailchimp email marketing software:
      https://executeyourtechidea.com/r/mailchimp

59   Microsoft affiliate program:
      https://executeyourtechidea.com/r/ms365-affiliate-program

60   The global affiliate marketing industry is now worth over $12 billion
      dollars and is expected to continue growing by around 10% over
      the next few years. Affiliate marketing statistics:
      https://executeyourtechidea.com/r/affiliate-marketing-stats

61   Such as Google and Facebook

62   Spotify premium plans:
      https://executeyourtechidea.com/r/spotify-premium

63   ASOS clothing e-commerce shop:
      https://executeyourtechidea.com/r/asos

64   ASOS sees their revenue reach over $3.26 billion:
      https://executeyourtechidea.com/r/asos-revenue

65   Big commerce online shopping statistics:
      https://executeyourtechidea.com/r/bigcommerce-stats

66   Mcdonalds franchising:
      https://executeyourtechidea.com/r/mc-franchising

67   Costa partner program:
      https://executeyourtechidea.com/r/costa-business

68   Toni and Guy franchising:
      https://executeyourtechidea.com/r/toniandguy

69   Amazon: https://executeyourtechidea.com/r/amazon

70   Shopify: https://executeyourtechidea.com/r/shopify

71   Notebook Therapy Website:
      https://executeyourtechidea.com/r/notebooktherapy
      Notebook Therapy on Instagram:
      https://executeyourtechidea.com/r/notebooktherapy-insta

72   Notebook Therapy on Trustpilot:
      https://executeyourtechidea.com/r/notebooktherapy-trustpilot

73   History of PAYG mobile phones:
      https://executeyourtechidea.com/r/payg-history

74   Google bought Android:
      https://executeyourtechidea.com/r/android-aquisition

75   Number of active smart devices running Android:
      https://executeyourtechidea.com/r/num-smartphones

76   I should call them Alphabet really, but you get the idea. Either way,
      here is data on their revenue: https://executeyourtechidea.com/r/
      alphabet-revenue

77   Play store revenue statistics:
      https://executeyourtechidea.com/r/app-revenues/

78   Amazon Web Services (AWS):
      https://executeyourtechidea.com/r/aws
      Asure, Google cloud and IBM are all alternatives to AWS.

79   For example, Highcharts, a company that makes it easier for
      developers to create interactive charts, licences their software
      per-developer that needs to use their tools
      https://executeyourtechidea.com/r/highcharts
      Non-commercial licence version of highcharts:
      https://executeyourtechidea.com/r/highcharts-noncomm

80   Uber taxi and private hire marketplace app:
      https://executeyourtechidea.com/r/uber

81   If your tech idea aims to be a marketplace business, the I highly
      recommend that you read the VersionOne guide to marketplaces:
      https://executeyourtechidea.com/r/marketplace-handbook

82   eBay online auction site and app:
      https://executeyourtechidea.com/r/ebay

83   eBay revenue growth stats:
https://executeyourtechidea.com/r/ebay-revenue

84   such as Manheim, SmartAuction, and LiveAuctioneers:
https://executeyourtechidea.com/r/manheim
https://executeyourtechidea.com/r/smart-auctions
https://executeyourtechidea.com/r/live-auctions

85   If you fancy a heavier read on the topic of business models, then I
recommend checking out *The Business Model Navigator* by Oliver
Gassmann and Karolin Frankenberger, who delve more deeply into
55 business models that can revolutionise businesses. They aren't
specifically geared around technology businesses, but they should
give you further inspiration.
https://executeyourtechidea.com/r/business-model-navigator

86   I hope you like Comic Sans, because that's what I've used for these
bad boys.

87   Norton rose fullbright – 2016 Litigation trends annual survey:
https://executeyourtechidea.com/r/litigation-trends
Klemm Analyst Group – Impact of litigation on small business:
https://executeyourtechidea.com/r/litigation-impact

88   Warwick university and IZA Bonn research by economists Andrew
J. Oswald, Eugenio Proto and Daniel Sgroi showed happiness made
people around 12% more productive on average, with companies
such as Google achieving productivity improvements of as much as
37% due to staff happiness levels:
https://executeyourtechidea.com/r/happy-study

89   Pryce-Jones research shows that the happy worker really is the
productive worker. Happy employees are 180% more energised,
155% happier with their jobs and life, 108% more engaged, and
50% more motivated too
https://executeyourtechidea.com/r/happy-productive

90   Gallup global health study shows unhappy employees take on
average 15 extra sick days per year when compared to happy
employees:
https://executeyourtechidea.com/r/global-health-study

91   Average revenue generated per employee in the UK:
https://executeyourtechidea.com/r/employee-revenue

92   "SMEs have written off £20,403 in bad debt over the past
      12 months, up 22%". Bibby financial services, SME confidence
      tracker report:
      https://executeyourtechidea.com/r/sme-confidence-tracker

93   Source: The Atradius Payment Practices Barometer - The
      Americas:
      https://executeyourtechidea.com/r/bad-debt-study

94   Gerald Ratner wipes £500m off of his company's value with a
      public speaking blunder, now known as the Ratner Effect:
      https://executeyourtechidea.com/r/gerald-ratner

95   It's 50% easier to sell to existing customers than to new prospects
      - "Marketing Metrics" by Paul W. Farris, Neil T. Bendle, Phillip E.
      Pfeifer, David J. Reibstein. ISBN:0137058292 9780137058297

96   This calculations assumes no customers unsubscribe, which is
      unlikely to happen in practice, but this is a simple example to prove
      the concept of how customer growth can impact annual revenue.

97   Research shows that just a 5% increase in customer retention can
      increase profitability by 25–75%, with repeat customers spending
      on average double between two and two-and-a-half years as a
      customer than they did within the first six months. The economics
      of E-Loyalty by Frederick F. Reichheld and Phil Schefter, Harvard
      Business School. https://executeyourtechidea.com/r/customer-
      retention-study

98   The proper accurate calculation uses "n * (n -1)" rather than "n *
      n", but I want to keep the maths as accessible as possible, n * n is
      close enough for an estimate, and is easier to remember.

99   It's 50% easier to sell to existing customers than to new prospects
      - "Marketing Metrics" by Paul W. Farris, Neil T. Bendle, Phillip E.
      Pfeifer, David J. Reibstein. ISBN:0137058292 9780137058297

100   In 2021, their brand was believed to be worth $87.6bn. Coca Cola
      Brand Value Estimates:
      https://executeyourtechidea.com/r/coca-cola-value

101   Sorry to all those Pepsi fans out there, as the same analogy applies
      to your favourite drink brands. And if you only drink water, then
      good for you!

102   Apple estimated brand value:
      https://executeyourtechidea.com/r/kantar-brand-value

103   If you remember from earlier, the customer lifetime value is
      the total amount of money we expect to make from an average
      customer based on how much they spend, how frequently, and for
      how long.

104   Return on investment.

105   This profit multiplier is usually called the "Price Earnings Ratio" in
      the investing world.

106   There are many different sources online where you can look up the
      average profit multipliers for each industry. For example: https://
      executeyourtechidea.com/r/sev

107   For example, in the UK, the Office for National Statistics provides a
      lot of population demographic data for free:
      https://executeyourtechidea.com/r/ons-data

108   Businesses are thought of as games in 'game theory': a maths-and-
      logic-based study of competitive game-like situations. In game
      theory, we have two main types of game: finite and infinite.
      https://executeyourtechidea.com/r/game-theory

109   Honourable mention to the fiction book *Player of Games* by Ian M
      Banks https://executeyourtechidea.com/r/player-of-games

110   Here is a link to my development agency Scorchsoft:
      https://executeyourtechidea.com/r/scorchsoft

111   Jira "You'll see value propositions and information about the value
      that JIRA brings all over the JIRA landing page:
      https://executeyourtechidea.com/r/jira

112   If you'd like to read more on why you should define personas in
      this way, you may want to check out the book *Designing for the
      Digital Age: How to Create Human-Centered Products and Services*
      by Kim Goodwin.

113   Aimee's website: https://executeyourtechidea.com/r/aimee

114   When you see photos of upcoming movies or shows in magazines
      and on social media, those aren't screen-grabs. An on-set

photographer took them, and depending on the show, it may have been Aimee.

115   Our platform is called ImageApprovals and has been running successfully for years now: https://executeyourtechidea.com/r/imageapprovals

116   In a real scenario, this list of features would be much longer and could be broken down into sub-features.

117   Going viral is where the number of people sharing something is so great that the total number people engaging with that thing grows exponentially. This is a similar effect to how a virus can spread, hence the term 'viral'.

118   Popular website builder platforms: Squarespace: https://executeyourtechidea.com/r/squarespace Wix: https://executeyourtechidea.com/r/wix Shopify: https://executeyourtechidea.com/r/shopify

119   Apple's success secrets. four key tactics of high NPS to the Apple brand: https://executeyourtechidea.com/r/apple-tactics

120   Including SoGoSurvey, Google Forms, and Typeform. Sogosurvey: https://executeyourtechidea.com/r/sogo Google forms: https://executeyourtechidea.com/r/google-forms Typeform: https://executeyourtechidea.com/r/typeform

121   Such as Google Sheets: https://executeyourtechidea.com/r/sheets and Google Forms: https://executeyourtechidea.com/r/google-forms

122   PBJUMPS jump training web app: https://www.scorchsoft.com/pbjumps-app-case-study

123   Google Trends: https://executeyourtechidea.com/r/google-trends and we also have many Google Trends alternatives: Semrush, Act-On, SE Ranking, Ahrefs, Serpstat, SpyFu, Moz Pro, and Bloomreach.

124   Such as AngelList: https://executeyourtechidea.com/r/angel-list

125   Credit checking companies include those like Creditsafe and DueDil. Some companies will let you see their profitability on their public accounts, but if they don't, you can get a rough idea by

looking at the reported "Shareholder funds". If the number is going up, this is an indicator that it's increasing because there have been profits. However, proceed with caution when using this metric, as it's common for very new companies to be unprofitable for many years after they start up. https://executeyourtechidea.com/r/credit-check , https://executeyourtechidea.com/r/duedil

126 Such as Campaign Monitor and Mailchimp: https://executeyourtechidea.com/r/campaign-monitor , https://executeyourtechidea.com/r/mailchimp

127 Although most software sales calls have the objective of scheduling the next phase of the sales process, usually a product demo as mentioned earlier.

128 Such as Cold Calling 2.0 techniques.

129 As with previous chapters I will separate maths or calculation examples in grey boxes so that they are easy to skim past for the casual reader.

130 Which came first, the chicken or the egg?

131 I won't disclose real sales and margin figures, so this example uses template numbers to illustrate the kind of challenges the venture faced.

132 I'll explain what a Venture Capital firm is a little later in this chapter.

133 VersionOne is a Venture Capital firm that specialises in investing in marketplace businesses. Here is their guide to marketplace businesses, which is essential reading if your business is a marketplace: https://executeyourtechidea.com/r/marketplace-handbook

134 At the time of writing, in the UK, there are government backed start-up loads that allow you to raise up to £25,000 (unsecured) per director at a reasonable interest rate. Here is how to apply for a government backed start-up loan for your UK business: https://executeyourtechidea.com/r/startup-loan

If you're reading this book from another country, then often your government or state may have similar schemes. Do an online search to see what opportunities exist where you live.

135 Do you need a cofounder? You might not:
https://executeyourtechidea.com/r/need-co-founder

136 You can get premium shareholders agreement templates online
to get a feel for their structure and content. I'd recommend you
download one of these and fill it out, then contact a solicitor to go
over it with you to make sure it's legally sound.

137 Based on what I've seen from investors in the UK, most individual
angel investors have the appetite to invest £10,000–£100,000 of
their own money. You can find investors looking for opportunities
larger than this, but they are rarer.

138 Startups in the UK can use the Startup Enterprise Investment
Scheme to raise money their business:
https://executeyourtechidea.com/r/seis
https://executeyourtechidea.com/r/seis-capital-gains

139 From what I've seen, most angel groups look to invest between
£50k–£300k. For amounts over £150k, they will normally combine
SEIS with another similar scheme called EIS. Use the Enterprise
Investment Scheme to raise money for your business:
https://executeyourtechidea.com/r/eis .

I appreciate that following all these different acronyms might
be challenging. I wouldn't get tied up with the jargon as it's not
important. All you need to focus on is that angels invest less
than angel groups, and there are tax reasons why.

140 In the UK, a typical SEED fund raising round aims to raise between
£300k and £1m.

141 Such as the Kickstarter crowd funding platform:
https://executeyourtechidea.com/r/kickstarter

142 Such as the Seedrs platform or CrowdCube.
https://executeyourtechidea.com/r/seedrs
https://executeyourtechidea.com/r/crowdcube

143 I've seen clients I've worked with raise around £300k with this
crowd funding method of finance, though the range is quite wide,
with the smallest campaigns starting around £10,000 and the
largest scaling up to multi-millions.

144   This is a very brief insight into the world of start-up funding, and I could write a book on this by itself. If you'd like to learn more about how to fund your idea, then I recommend checking out *The Startup Funding Book* by Nicolaj Nielsen, which goes into much more depth about the different ways to raise capital, giving practical examples on what steps to take, including creating a pitch deck and how to reach out to prospective investors.

145   *Crossing the Chasm* by Geoffrey Moore:
https://executeyourtechidea.com/r/cross-chasm

146   *What Makes for Successful Innovation Teams in Small and Medium Enterprises? A Multiple Case Study*:
https://executeyourtechidea.com/r/successful-innovation

147   I know I've used Apple and Musk a lot in this book, and it is because they are both tech companies that a lot of people recognise. It also means I can build a picture using lots of slightly different examples using the same companies.

148   Manager Tools guidance on the three types of power:
https://executeyourtechidea.com/r/three-types-of-power

149   As I mentioned earlier, I'm following the Pareto 80/20, where the approach you take is the 20% effort that delivers the 80% of the value.

150   Paint a picture of the dog onto a canvas, not apply paint to a dog.

151   Here, I assume you will follow a Waterfall project approach as this is a common method used to release the first version of a project. However, if an Agile project management style suits you better, then the way you document and plan your project will be different. I'll cover the difference between Waterfall and Agile in the Project Lifecycle Types Chapter and will expand on Agile further near the end of the book in the Plan Your Product Roadmap chapter.

152   Balsamiq is an excellent easy-to-use tool for beginners to create wireframes:
https://executeyourtechidea.com/r/wireframetool

153   Code Academy has free and paid course content. The free content is enough to learn the basics and the paid content will help develop

your skills further if you think programming is for you.
https://executeyourtechidea.com/r/codeacademy

154 Code with Mosh is one of my favourite sources to learn
development. His videos are super clear and concise – no waffle,
and his online courses are very well-structured.
https://executeyourtechidea.com/r/moshchannel
https://executeyourtechidea.com/r/moshsite

155 Kahn Academy is arguably the best free education resource in the
world and covers a huge range of different topics across an array
of different of disciplines, including programming.
https://executeyourtechidea.com/r/khan

156 Reddit r/learnprogramming : The Learn Programming subreddit
contains tons of resources around how to learn to code, some
aimed at adults and others at children, such as sites that gamify
learning to code. I would recommend reading the wiki on this
subreddit to see if any of the links or advice resonate with you:
https://executeyourtechidea.com/r/learnprogramming

157 Squarespace – Website builder:
https://executeyourtechidea.com/r/squarespace

Shopify – Ecommerce builder:
https://executeyourtechidea.com/r/shopify

Jotform – Online form builder:
https://executeyourtechidea.com/r/jotform

Google Forms – Online form builder:
https://executeyourtechidea.com/r/gforms

iAuditor - health and safety app maker:
https://executeyourtechidea.com/r/auditgen

Firebase – DIY -back-end infrastructure:
https://executeyourtechidea.com/r/firebase

IBM Watson – Artificial Intelligence platform:
https://executeyourtechidea.com/r/watson

Airtable – Interactive spreadsheet apps:
https://executeyourtechidea.com/r/airtable

Microsoft Asure – Cloud computing solutions:
https://executeyourtechidea.com/r/azure

Google Cloud solutions: https://executeyourtechidea.com/r/gcloud

Amazon Web Services: https://executeyourtechidea.com/r/aws

158   Parse failed: https://executeyourtechidea.com/r/parse-failed

159   Though I should add, there is now an open source project which aims to keep some of the value first created by the Parse platform aimed at developers who are happy to install that platform onto their own servers https://executeyourtechidea.com/r/parse-open-source

160   We covered what an API was in the "The Tech Supporting your Idea" chapter. But as a refresher, API stands for Application Programming Interface. If a system has one, it's a way that other systems can automatically communicate with it.

161   Indonesia salaries: https://executeyourtechidea.com/r/indonesia-salary

162   Swiss average Salary: https://executeyourtechidea.com/r/swiss-salary

163   Which Five Companies Do The Most Overseas Manufacturing? https://executeyourtechidea.com/r/most-manufacturing

164   Bangladesh factory collapse toll passes 1,000: https://executeyourtechidea.com/r/factory-collapse

165   Data from The WOW Company survey of 471 agencies in 2017: https://executeyourtechidea.com/r/avg-agency

166   Other tech experts in the industry might disagree with my opinion of the usefulness of tenders to source suppliers for tech projects.

167   You might also find that some of them have written entire books on the process of executing your tech idea!

168   These too have different specialisms, such as back end, front end, Devops, and full stack, which goes beyond the scope of this book.

169   Epitomise is one example of a marketing consultant company: https://executeyourtechidea.com/r/epitomise

170   For example, if a company only has five staff, then how can
       you have design, development, project management, testing,
       marketing, and sales? In this scenario, the chances are that each
       person will wear many hats. Is it possible for a developer to be
       good at design, development, and marketing? I'll let you decide.

171   We covered different elements of the technology stack in
       "Chapter 2: The Tech Supporting Your Idea".

172   Unless you already have a tool that you must continue to support.
       In this case, it may be sensible to source a developer to support
       the existing tool who has experience programming in the same
       languages it was originally built with.

173   This "perpetual, royalty-free licence to use and extend the code"
       is an example legal wording for unrestrictive licence terms from
       my experience. Please remember I am not a legal professional so
       if you are unsure about the rights you have, then I recommend
       engaging a solicitor, even if it's just to check over the IP clauses in
       the contract.

174   The amount to spend on testing and project management will
       depend on the nature of your project. Some projects are simple
       and require fewer resources to manage and test. Others can be
       very technically complex requiring high amounts of weekly activity.

175   I add this question with caution as skipping elements of the design
       process can save on costs but be a serious factor in reducing
       the quality of your project. However, I also accept that a highly
       polished design might not be needed for prototypes or internal
       tools used by a very small number of people. In this case, it may be
       possible for the development team to do an OK job by coding the
       UI of the platform guided by just the wireframes or a description of
       the requirement.

176   Good rapport is often how well you get along with the key contact.
       Do you feel a connection to them? Do you feel like they're helpful?
       Is there respect on both sides? Does communication flow easily
       between you both?

177   UML stands for Unified Modeling Language and is a set of
       commonly used ways to visually describe the requirements of a
       software development project with diagrams. It's down to your

tech team to decide which of these they think is most important for your project.
https://executeyourtechidea.com/r/uml-examples

178   For small projects that require 100–300 development and project management hours to produce, then a one–two hour initial planning meeting followed by one or two 60–minute follow-up calls is typically enough to create the wireframes, specification, and other supporting information.

179   World leaders in research-based user experience, the Nielsen-Normal Group, conducted research that showed significant improvements in project success (or failure avoidance) for projects that conduct a discovery phase:
https://executeyourtechidea.com/r/nn-discoveries

180   There are many diagrams that exist to map out the characteristics of your project called UML diagrams. Depending on the size, complexity, and nature of your project, your chosen tech team may want to supplement the specification with these varied resources.

181   If you'd like to go one step further than sending your scope to suppliers, you can create a "Request for Quote" (RFQ) document instead. This is very similar to the scope but includes other information about your business needs, such as a background and objectives, expectations of the chosen supplier relationship, budget range, expectations around timescales, how you will measure success, or any other supporting resources or information.

182   Such as Jira or Asana.
Jira: https://executeyourtechidea.com/r/jira
Asana: https://executeyourtechidea.com/r/asana

183   Agile, as I've described it in this book, is just a stepping stone towards working in a fully Agile way. Going fully Agile requires you to follow the Agile Manifesto in full and may not always be a process or culture fit for your business.
https://executeyourtechidea.com/r/agile-manifesto

184   There are many types of Agile project management approaches: Scrum, Kanban, Lean, Extreme Programming, Scrumban, and Feature-Driven Development (FDD) to name a few. I've only touched the surface of these two common project management

methods so you know what to expect as your project is built and launched. Each of these Agile methods is so extensive that you will find dedicated books on them. Here are the top 15 Agile books you should read if you want to learn the difference between the various Agile ways of working:
https://executeyourtechidea.com/r/top-agile-books

185 Epitomise strategic marketing, my father's company:
https://executeyourtechidea.com/r/epitomise

186 Source, Harvard Business Review "The Value of Keeping the Right Customers": https://executeyourtechidea.com/r/five-times

187 Why it Takes 6-8 Marketing Touches To Generate a Viable Sales Lead: https://executeyourtechidea.com/r/8-touch-sale

188 Organic means any form of digital marketing that does not involve paid ads.

189 Growth may be a goal, but if you grow before you're ready, then you risk multiplying any business inefficiencies or challenges you have. Sometimes it can be better to ramp up slowly so you have time to adjust in a controlled way.

190 A notable mention to the book *24 Assets* by Daniel Priestly which outlines an extensive list of 24 types of digital asset that organisations can invest to create to drive long-term value:
https://executeyourtechidea.com/r/24-assets

191 Examples of popular search engine include Google and Microsoft's Bing.

192 This model booking example was first mentioned in 'Chapter 8: Raising and Managing Money'

193 Google Keyword Planner:
https://executeyourtechidea.com/r/keyword-planner
Bing Keyword Research tool:
https://executeyourtechidea.com/r/bing-research-tool

194 Google and Bing also offer this kind of search ad, as do other providers.

195 The example uses Google Merchant Centre and Shopping:
https://executeyourtechidea.com/r/merchant-center

196 Organic Search Engine Optimisation (SEO) is the process of trying to get your website pages to come up organically on SERPS. It involves creating high-quality pages, then doing activities to try to get those pages to perform. This is a diverse and skilled field, and can often involve many days of reaching out to sites and sources to get quality websites to link back to your pages. If you want to learn more about SEO, then I'd recommend you check out the Neil Patel blog: https://executeyourtechidea.com/r/neilblog
And the Backlinko blog:
https://executeyourtechidea.com/r/backlinko

197 Source for the click and conversion benchmarks:
https://executeyourtechidea.com/r/adwords-benchmarks

198 Source on positioning on page relating to results:
https://executeyourtechidea.com/r/google-ctr-stats

199 Purchasable well-known tools include SEMRush, Moz, and Ahrefs.

200 Google search console:
https://executeyourtechidea.com/r/search-console

201 Keyword Planner:
https://executeyourtechidea.com/r/keyword-planner

202 Google Ads: https://executeyourtechidea.com/r/g-ads

203 Google My Business:
https://executeyourtechidea.com/r/g-my-biz

204 Google Analytics: https://executeyourtechidea.com/r/g-analytics

205 PageSpeed Insights:
https://executeyourtechidea.com/r/page-speed

206 Trends: https://executeyourtechidea.com/r/google-trends

207 Data Studio: https://executeyourtechidea.com/r/data-studio

208 Bing Webmaster Tools:
https://executeyourtechidea.com/r/bing-webmaster

209 An example of a software comparison site that also provides online

reviews is Capterra: https://executeyourtechidea.com/r/capterra

210 Examples of procurement frameworks include the G-Cloud (Cloud Services), DOS (Digital Outcomes and Specialists), and Crown Commercial Services frameworks.

211 The digital marketplace is used by public sector organisations to find cloud technology and specialist services for digital projects (https://executeyourtechidea.com/r/digigov) and if approved, you will be added to the supplier list (https://executeyourtechidea.com/r/govcloud).

212 B2B Buying Journeys May Stretch 12 Months Or Beyond, But Most Key Decisions Are Made In The First 90 Days: https://executeyourtechidea.com/r/b2bjourneys

213 (Business Network International), British Chambers of Commerce, Federation of Small Businesses (FSB), and Institute of Directors (IOD) are just four examples of extremely popular UK business networking groups. The Entrepreneur Handbook provides an overview of these and others. https://executeyourtechidea.com/r/business-networks

214 Two example sites include
Tender Rocket: https://executeyourtechidea.com/r/tenderrocket
Find a Tender: https://executeyourtechidea.com/r/gov-tender

215 Department for International Trade: https://executeyourtechidea.com/r/dit

216 An example is Tender Rocket: https://executeyourtechidea.com/r/how-to-apply-for-tenders

217 Such as Computer Weekly, TechRadar, ZDNet.

218 Such as Forbes, Insider, Entrepreneur, and so on.

219 Such as Gartner, IDC, Forrester Research, and Frost and Sullivan, who are some of the larger established tech industry experts, analysts, and consultants.

220 Later, we look at influencer marketing as an outbound marketing channel, but due to the high number of followers that influencers typically have in their networks, they can equally be used by purchasers searching for opinions and recommendations.

221 You can display your message on the Google Display Network (GDN), on Bing's display network, on social media networks (YouTube and partners, LinkedIn, Facebook, Instagram, Twitter, etc.), or on trade or industry-specific websites.

222 Such as the Telegraph or the Guardian if you are a UK reader.

223 At the time of writing, this 0.39% figure is for Google display advertising.

224 Google Ads Benchmarks for various Industries: https://executeyourtechidea.com/r/adwordsbenchmarks

225 Case studies illustrating how various companies have used different elements of the Google Marketing Suite (including display ads) can be viewed on their 'Think with Google' website: Case studies: How brands are innovating on YouTube: https://executeyourtechidea.com/r/brand-marketing-case-studies

226 Examples of mainstream platforms include Facebook, YouTube, WhatsApp, Facebook Messenger, LinkedIn, Twitter, TikTok, WeChat, Reddit, Tumblr, Instagram, Snapchat, Pinterest, YouTube, Facebook, Periscope, Vimeo, and Reddit. There are also many niche platforms with examples including About.me, Periscope, Inbound.org, and beBee.com.

227 Statista is an example: https://executeyourtechidea.com/r/statista

228 These are hypothetical numbers.

229 In the UK alone, there are 48 million active social media users, with the average spending 110 minutes on social media per day: UK Social Media Statistics and Facts: https://executeyourtechidea.com/r/social-media-statistics-uk

230 There are many tools to support productive automated posting, monitoring, and engagement on social media, and most businesses will use one or more of these. Examples include HootSuite, HopperHQ (for Instagram), Postfity, Sprout Social, Buffer, Feedly, SocialBee, and ContentCal.

231 Such as LinkedIn Sales Navigator. We're not sponsored by these recommendations, I promise! I recommend them as they are tools I use regularly to deliver success for my clients.

232 There are many website visitor identification tools, often referred to as lead management software tools, including Lead Feeder, VisitorQueue, Lead Forensics, and LeadInfo.

233 Especially if those databases are regulatory-compliant, including GDPR for European audiences, with explicit opt-in permission to contact.

234 Mailchimp, Campaign Monitor, Pure360, HubSpot and Active Campaign, Marketo, and Oracle are examples of MAPs and email systems. A valuable comparison of these platforms is available from Capterra: https://executeyourtechidea.com/r/capterra-software

235 The latest DMA Marketer email tracker shows that marketers rated email as the most strategically important communications channel, with 91% of those surveyed rating it as 'important'. https://executeyourtechidea.com/r/email-marketing-trends

236 Source, industry marketing benchmarks: https://executeyourtechidea.com/r/email-marketing-benchmarks

237 10 Webinar Benchmarks Every Marketer Should Know: https://executeyourtechidea.com/r/webinar-benchmarks

238 According to goto.com, 29% of all webinars are technology and software companies: Audio on demand, the rise of podcasts, and the annual growth rate has been over 24%: https://executeyourtechidea.com/r/ofcom-podcasts

239 The need-feature-benefit analysis was first introduced in Chapter 7: Your Minimum Viable Product (MVP).

240 However, when sending such items, it's important to ensure compliance with relevant bribery and corruption acts, which in the UK is the Bribery Act 2010. Many big companies require employees to reject gifts that are over £50 in value.

241 The Direct Marketing Association state that the average response rate to a direct mail campaign is 9%, and this includes the typical consumer direct mail pieces too: https://executeyourtechidea.com/r/direct-mail-response-rate

242 The 'Predictable Revenue' Cold Calling 2.0 technique recommends that you split prospecting and sales functions to achieve maximum results: https://executeyourtechidea.com/r/predictable-revenue

243 100 calls may seem high, but it's common in some environments that telemarketers are expected to call 80 to 100 people each day.

244 If affiliate marketing is prioritised as one of the main channels for your business, Neil Patel's blog 'What is Affiliate Marketing' describes how to get started as a 'merchant' seeking affiliates for your new product: https://executeyourtechidea.com/r/neilpatel

245 Check out the Google Campaign URL Builder to add campaign information to URLs. You can make a pretty URL that redirects to a less pretty campaign URL: https://executeyourtechidea.com/r/utm-builder

246 For guidance on ways you can become a key person of influence in your niche, I highly recommend the *Key Person of Influence* book by Daniel Priestly: https://executeyourtechidea.com/r/key-person-of-influence

247 Fun fact: this advice was given to Andrew in person by Jujimufu on a call some years ago. Juji is an influencer in the health and fitness space that Andrew knows from his days doing acrobatics and tricking, and he has a YouTube following of over 1.3 million subscribers at the time of writing. Andrew's pet cat is also called Juji. Watch Jujimufu on his YouTube channel here to see an relevant influencer example: https://executeyourtechidea.com/r/jujimufu

248 Word-of-mouth marketing: https://executeyourtechidea.com/r/wom-marketing

249 Sourced from the paper 'The Impact of Celebrity Endorsement on Consumer Buying Behavior' by Humaira Mansoor and Muhammad Mehtab Qureshi: https://executeyourtechidea.com/r/negative-celeb

250 Richard Thaler covers the concept of celebrity endorsement efficacy in his book on behavioural economics, *Misbehaving*: https://executeyourtechidea.com/r/misbehaving-book

251 Across the UK, there are currently over 11,300 outdoor digital screens, and the total spend on out-of-home advertising in the

UK was over £700m in 2020, which has a predicted post-Covid growth rate of 36%. Source: Out-of-home advertising in the United Kingdom (UK) — statistics & facts:
https://executeyourtechidea.com/r/out-of-home-advertising

252 One unforeseen benefit of printed poster advertising is that even though you've paid for your poster to be displayed for a certain window of time, sometimes they can be slow to take down or replace your poster once the advertising period has ended. You can't plan for this, but if you do enough billboard and poster advertising, then sometimes you may get lucky, especially for advertising in unusual, niche locations.

253 Though the billboard ad providers do try, it's hard to verify that what they say is an accurate model. They may calculate their estimates based on several rough factors, which give them a number to tell you that looks good.

254 Superbowl viewing data
https://executeyourtechidea.com/r/superbowl

255 Marketing Donut's blog, titled "Could Your Business Afford to Advertise on TV?" provides examples of TV advertising pricing, this ranges from under a hundred to many thousands of pounds depending on the channel and time.
Source: "Could your business afford to advertise on TV?"
https://executeyourtechidea.com/r/tv-afford

256 The engineer who thinks the company's AI has come to life:
https://executeyourtechidea.com/r/ai-sentient

257 According to Zendesk, interactions with chatbots grew 81% in 2020: What is a chatbot? Types of chatbots and how they work:
https://executeyourtechidea.com/r/chatbots-for-business

258 Chatbots are growing fast, and are worth tens to hundreds of billions in spend:
https://executeyourtechidea.com/r/chatbot-market-stats

259 of Open Innovation study suggests that chatbots can increase leads by 25%:
https://executeyourtechidea.com/r/chatbot-implement-study

260 Comparing Data from Chatbot and Web Surveys: https://executeyourtechidea.com/r/chatbot-surveys

261 Many guerrilla marketing campaigns are highly visual and the Word Stream blog, called 20+ Jaw-Dropping Guerrilla Marketing Examples', illustrates: https://executeyourtechidea.com/r/guerrilla-marketing-examples

262 You can view the full colour of this NHS Guerrilla Marketing image version here: https://executeyourtechidea.com/r/nhs

263 I appreciate there is quite a lot of information in this image, so here's a link to a larger version should you want to zoom in and out more closely: https://executeyourtechidea.com/r/marketing-impact-effort

264 If you wish to go deeper, there are also digital tools that can be used to obtain a view of where any website is getting its visitors from. For example, what organic and paid search terms they're using and where their display ads are being shown. Understanding this helps you form a judgement of what is working for others, and this can form part of your consideration of the right communication channels for your business and its new product.

265 For example, you can quickly set up the free tool Google Data Studio to provide a visual view of your funnel, combining data from other sources into a single view dashboard. These other sources may include data from Google Analytics, Google Search Console, Google Ads, Bing webmaster measure tools, all of which are free.

266 If you aren't familiar with these tools, YouTube has many videos showing an overview of them and instructions to help you set them up if you want to do this yourself versus using an agency. The key to delivering the desired results is knowing where you're at, doubling down on what's working, and adjusting or stopping what isn't. Having a measurement environment will help you achieve this.

267 Small Businesses Seeking Higher Growth Regret Not Investing In Marketing Sooner: https://executeyourtechidea.com/r/invest-in-marketing-sooner

268 Facebook could have been called Facemash. When did Facebook start? https://executeyourtechidea.com/r/fb-start

Mark Zuckerberg Bloomberg profile: https://executeyourtechidea.com/r/zuckerberg

269 In my opinion, this is a major reason why so many public sector technology projects go over budget whilst simultaneously failing to deliver the value originally promised. They were denied the ability to pivot or change direction as priorities changed. Especially when contracts are awarded on seven-year renewal cycles as it's inevitable that requirements will change over such a long time.

270 Don't forget that the trusty impact-effort matrix can help you to order these competing priorities.

271 Here are some examples of popular project management tools:
Atlassian Jira: https://executeyourtechidea.com/r/jira
Asana: https://executeyourtechidea.com/r/asana
Aha! https://executeyourtechidea.com/r/aha
ProductPlan: https://executeyourtechidea.com/r/productplan

272 The ticket workflow will also vary significantly depending on the working practices of your team, for example, teams using a 'Git Flow' technique will follow different deployment processes to those who like 'Continuous Integration'. These topics are well outside the scope of this book, but you should know that there are good reasons for tech teams to have different preferences.

273 Here is a large resolution version of the ticket flow should you want to zoom in and out on your device:
https://executeyourtechidea.com/r/ticket-flow

274 Git and SVN are two common types of version control tool. Git is by far the most popular.

275 The OWASP Zed Attack Proxy is one example of an open source penetration testing tool that can be used to check the security of your tech: https://executeyourtechidea.com/r/zap

276 The better written and better structured the code is, the fewer comments it may require. In practice, what the code does may not always be clear, and explaining the 'why' behind a coding decision is necessary. My preference is to always comment classes and functions even if the comments are minimal. Have small enough functions that the function name explains the function purpose, then add comments inside functions where the reasoning behind a

certain decision may not be obvious. It's also good practice to add comments where code may not be inherently easy to understand by just reading it, such as when your developer has written something particularly clever.

277 Vertical scaling is where you upgrade the existing servers that code runs on to be more powerful. Horizontal scaling is where you make your service run across more servers or web services to cope with the increased load. Vertical scaling is usually cheaper and less complex than horizontal scaling.

278 Scorchsoft: https://executeyourtechidea.com/r/scorchsoft

279 Trade.gov article on ICT opportunities in the UK: https://executeyourtechidea.com/r/trade-uk-stats

280 Andrew's personal website: https://executeyourtechidea.com/r/andrew

281 Andrew Competes in the 74kg Open category in the International Powerlifting Federation (IPF). Here is a link to Andrew's Open Powerlifting page where you can see his competition performances: https://executeyourtechidea.com/r/open-powerlifting

282 The 100 most influential B2B tech marketers in Europe: https://executeyourtechidea.com/r/marketing-leader